Industrial Processes and Waste Stream Management

Industrial Processes and Waste Stream Management

Editor

Vimal Kumar

scitus
academics

Industrial Processes and Waste Stream Management
Edited by **Vimal Kumar**

Printed in 2017

ISBN: 978-1-68117-405-1

Library of Congress Control Number: 2015941602

© 2016 by
SCITUS Academics LLC,
616, Corporate Way, Suite 2, 4766,
Valley Cottage, NY 10989

www.scitusacademics.com

Contents

Preface

Waste management is largely regulated by legislation and policy implemented at the municipal level, but there are significant provincial regulations that may come into play. In some instances federal regulations may also be relevant, particularly if dealing with hazardous substances or shipping waste across provincial boundaries. Waste diversion and waste minimization are often a primary focus for most integrated waste management plans. Specific goals and targets are defined in a plan. In many jurisdictions, the ICI sector must follow prescribed federal, provincial and municipal goals and targets as identified in acts, regulations, and bylaws. The book lays the conceptual foundations with a detailed overview of waste stream management tools and regulations and the four EPA-approved treatment methods: physical, chemical, thermal, and biological. And provide a fascinating case-by-case exploration of industrial processes and how the waste streams they generate are managed in all major industries, including petroleum, chemicals, mining, metals, paint, textiles, agriculture, paper, printing, nuclear, medical, and more.

Editor

Third-generation Feed Stocks for the Clean and Sustainable Biotechnological Production of Bulk Chemicals: Synthesis of 2-hydroxyisobutyric Acid

Denise Przybylski, Thore Rohwerder, Hauke Harms, and Roland H Mueller

UFZ–Helmholtz Centre for Environmental Research, Permoserstrasse 15, Leipzig, 04318, Germany

ABSTRACT

Background

The synthesis of 2-hydroxyisobutyric acid (2-HIB), a promising building block for, e.g., Plexiglas® production, is described as an example for a clean and sustainable bioproduction.

Methods

A derivative strain of *Cupriavidus necator* H16, impaired in the poly-ß-hydroxybutyrate synthesis pathway and equipped with xenogenic 2-hydroxyisobutyryl-coenzyme A mutase from *Aquincola tertiaricarbonis* L108, was applied. Batch cultivation was performed in the presence of vitamin B12 by supplying a gas mixture comprising hydrogen, oxygen, and carbon dioxide.

Results

Exploiting the chemo-litho-autotrophic potential of this so-called knallgas bacterium, 2-HIB was synthesized and excreted into the cultivation broth under aerobic conditions when inorganic nitrogen-limited conditions allowed an overflow metabolism of carbon metabolites. 2-HIB synthesis proceeded at a rate of 8.58 mg/[(g bacterial dry mass)·h]. Approximately 400 mg/L in total was obtained. The results were subsequently compared to calculated model data to evaluate the efficiency of the conversion of the substrates into the product. To achieve overall yield data regarding the substrate conversion, the model describes an integral process which includes both 2-HIB synthesis and biomass formation.

Conclusions

This study has confirmed the feasibility of the microbial synthesis of the bulk chemical 2-HIB from hydrogen and carbon dioxide by exploiting the chemo-litho-autotrophic metabolism of *C. necator* H16 PHB⁻4, additionally expressing the foreign 2-HIB-coenzyme A mutase. The product synthesis was satisfying as a proof of principle but does not yet approach the maximum value as derived from the model data. Furthermore, the biosynthesis potential of an optimized process is discussed in view of its technical application.

BACKGROUND

A growing global population and rising living standards inevitably enforce the conflict between satisfying the people's demands for goods and services, on one hand, and the sustainable development requirements and the considerate treatment of nature and earth's resources, on the other. It is not only that fossil carbon sources will become limited in the future, but there is also a growing pressure to renounce the exploitation of currently treated and prospected sites for environmental reasons. Accidents, such as the recent oil disaster in the Gulf of Mexico, motivate this tendency even more. To stop the rigorous and reckless exploitation of the earth's resources, alternative resources must be recovered, and clean techniques have to be developed, offered, and applied. The turnaround in thinking and acting has been already evident in recent times, mostly with respect to energy production, for which sustainable resources and clean techniques are increasingly implemented to substitute today's oil and coal-based production [1].

Likewise to the issue of clean energy production, a change is necessary in the societies' approach of how to improve the future production of commodities [2-4]. Thereby, industrial (white/green) biotechnology offers an elegant way to provide alternatives [5-7] via the application of microorganisms or components of cells in combination with a broad spectrum of new-generation renewable substrates.

Applying biotechnological processes, the chemical industry has for ages produced, for instance, alcohols and organic acids mainly to be employed as chemicals, but above all as energy carriers in bulk quantities. Of those, bioethanol [8-11] is a recent example of modern fuels for motor vehicles[12], while biobutanol is expected to be another one [13,14]. Based on this knowledge of how to produce bulk-scale energy carriers, recent intentions envisage the extension of platform chemicals for wider applications [13,15-18]. Special attempts are directed towards the synthesis of chemicals, such as 1,3-propanediol [19], succinate [20], gluconic acid [21-23], or citric acid [24]. Likewise, 2-hydroxyisobutyric acid (2-HIB) fits well into this scheme as it is gaining importance as a platform chemical. In particular, it can be used as a precursor for methacrylic acid [15,25], a monomeric compound required for the synthesis of such prominent products like Plexiglas® (Evonik Röhm GmbH, Essen, Germany) and as an important ingredient for coating materials, paint, and glues.

In general, traditional biotechnological processes, such as those for bioethanol production, are based on carbon sources of the so-called first generation, i.e., carbohydrates such as sugars or starches directly derived from plants [26]. Nowadays, the focus has shifted to second-generation feed stocks[11,27], which rely on complex plant materials, such as cellulose, hemicellulose, and lignin; the monomers of which are, however, more difficult to access [28], especially in terms of biosynthesis. Additionally, the concept of using arable land to grow plants solely as substrate resources for biotechnological processes is a matter of controversial debate, not only in view of substituting natural biotopes (e.g., rain forests) by monocultures [29], but also in view of reasonable conflicts with nutritional issues and the food production industry [30,31].

One solution and actual expectation in terms of a future sustainable bulk chemical production is the utilization of substrates of the third generation, i.e., diverse gas mixtures which deliver carbon as well as reducing power from different sources [4,32,33]. This implies the utilization of CO_2 as a carbon source since CO_2 accumulates as a waste product of energy production from fossil resources. At the same time, the resulting consumption of CO_2 within such a new production scheme also provides a fundamental argument to support processes that counteract climate change

[13,34]. The required reducing power might be delivered by hydrogen generated, e.g., by solar energy [35] or wind power[36]. Some perspectives of how to use CO_2, not only as a substrate for diverse syntheses, but also for various biosyntheses, have recently been presented at the *Dechema colloquium*[37,38]. Among others, the processes developed by Coskata Inc., Illinois, USA, employing a variety of materials which can be converted into renewable fuels and chemicals by biofermentation of synthesis gas, have been demonstrated. Also, the development of special designer bugs, being capable of using flue gas as a substrate, has been introduced at the colloquium *Sustainable Bioeconomy*[39]. However, the biggest challenge of those miscellaneous approaches is and will be the competition with the established processes and the implemented production schemes of the chemical industry [3,31,40], where the biobased synthesis is often still defeated. Nevertheless, in this investigation, another perspective of how to use CO_2 to sustainably produce 2-HIB as a building block is presented.

We recently discovered a novel enzyme, the 2-HIB-coenzyme A mutase, which proves to be an ideal catalyst for the production of 2-HIB, especially, given that 2-HIB synthesis with this enzyme only requires a one-step isomerization of metabolites that are essential for the metabolism of a wide range of bacteria, i.e., 3-hydroxybutyryl-coenzyme A (3-HB-CoA) [25,41-44]. The synthesis of 2-HIB and its excretion into the cultivation broth can be realized by employing strains that express this heterologous enzyme in combination with an existing overflow carbon metabolism. The selection of suitable strains thus allows different substrates for the production of 2-HIB to be utilized, as has been demonstrated by using fructose [45], D Przybylski, unpublished work]. However, in seeking sustainability, the application of fructose, a substrate of the first generation, will not meet the requirements to qualify carbohydrates as future substrates.

Therefore, we have applied the 2-HIB-coenzyme A mutase to demonstrate the sustainable and clean production of 2-HIB from carbon dioxide and hydrogen by exploiting the chemo-litho-autotrophic metabolism of the knallgas bacterium *Cupriavidus necator* (*Alcaligenes eutrophus*) H16 PHB⁻4[46,47]. The synthesis of 2-HIB was successful at the experimental proof of principle stage. Model data were added to confirm the metabolic potential of such a process.

METHODS

Bacterial Strains and Plasmids

C. necator, strain H16 PHB⁻4 DSM 541 [47], was obtained from the DSMZ (Leibniz-Institut DSMZ - Deutsche Sammlung von Mikroorganismen und Zellkulturen GmbH, Braunschweig, Germany) and modified by introducing the plasmid pBBR1MCS-2::HCM [48], which originates from the broad-host-range cloning vector pBBR1MCS [49]. The plasmid contains the genes *hcmA* and *hcmB* coding for the two subunits of the 2-hydroxy-isobutyryl-coenzyme A mutase from *Aquincola tertiaricarbonis* L108 [41,44]. The plasmid was kindly provided by Evonik Industries AG (Marl, Germany).

Cultivation Conditions

The general cultivation was performed in Luria Bertani broth (Miller) at 30°C, and the strain was stored on LB-agar plates at 4°C. For batch cultivations, a mineral salt medium was used, as described by Schlegel and co-authors [50] supplied with 0.3 mg/L kanamycin and 50 mg/L vitamin B12. The pre-cultures were prepared from single colonies at 30°C and 150 rounds per minute (rpm) in 200 mL of the same medium with fructose as the sole carbon source under aerobic conditions. After fructose exhaustion, the pre-culture was used to inoculate a fresh culture which was immediately shifted to hydrogen and carbon dioxide. The cultivation continued in a batchwise manner under laboratory conditions at 22°C, using a shake flask equipped with a stirrer and containing a working volume of 0.6 L, gassed with a sterile mixture of $H_2:O_2:CO_2$ in variable ratios. Agitation was set to 200 rpm. The two gases apart from oxygen were supplied from a storage tank with a volume of 18 L treated according to the gasometer principle. The initial gas concentrations were about 25% to 50% H_2, 15% to 30% CO_2, and 10% to 20% O_2. The gases were supplied to the culture by a hollow fiber module (Fresenius, St. Wendel, Germany), using a membrane pump at a feeding rate of 750 mL/min moving a gas circuit. Hollow fibers had a pore width of 0.2 µm and a specific exchange area of 0.7 m^2. The external volume of the hollow fiber module was flushed with the bacterial suspension at a rate of 42.6 L/h, fed with a gear pump out of the shake flask. After passage through the module, the gases and the suspension were collected in the flask and separated from each other. The gases were recirculated to the gas tank and mixed with the residing gases by a propeller by means of a magnet-coupled motor installed outside of the tank, whereas the suspension was re-fed to the module. The consumption of gases was monitored both in terms of the change of the total volume, which was registered by the horizontal movement of the gas tank, and in terms of the concentration measured by three specific sensors. If required, specific gases were refilled into the gas tank. As there was no automated pH control in this simplified cultivation system, the pH was monitored off-line and adjusted to pH 7.0 by adding the required volumes of 10% NaOH according to a titration curve based on the growth medium.

On-line Analysis

The gas concentrations were measured by specific sensors for hydrogen (0% to 100%), oxygen (0% to 100%), and carbon dioxide (0% to 50%) (BlueSens, Herten, Germany) and were continuously monitored.

Off-line Analysis

The biomass concentration was monitored by the optical density at 700 nm (U-2000 Spectrophotometer, Hitachi High-Technologies Corporation, Tokyo, Japan) and converted into bacterial dry mass according to a calibration curve prepared earlier. The substrate consumption and 2-HIB synthesis were analyzed by isocratic HPLC (Shimadzu Corporation, Kyoto, Japan) using a Nucleogel Ion 300 OA column (300 × 7.8 mm, Macherey-Nagel GmbH & Co. KG, Düren, Germany) at 70°C with 0.6 mL/min 0.01 N H_2SO_4 as the eluant.

Evaluation Methods

The gas consumption was calculated from the differential changes of the total and individual gas concentrations by means of simple linear regression for the different phases of the fermentation.

Calculations

3-Phosphoglycerate (PGA) was defined as the central carbon precursor [51,52] from which the complete biomass synthesis was derived. The molar composition of the biomass in the model was taken as $C_4H_8O_2N$. It is synthesized from adenosine triphosphate (ATP) as the general energy carrier and proceeds with an efficiency of 10.5 g bacterial dry mass pro mol ATP [53]. The overall balance equation for the biomass synthesis from PGA is as follows:

$$4\,PGA + 29.1\,ATP + 3\,NH_3 + 5.5\,[2H] \rightarrow 3\,C_4H_8O_2N + 10\,H_2O \qquad (1)$$

[2H] denotes the reduction equivalents, which in general correspond to reduced nicotinamide adenine dinucleotide (phosphate) {NAD(P)H + H$^+$}.

RESULTS AND DISCUSSION

Theoretical product yields

To define the possible product yields in a growth-associated process, we applied a stoichiometric model. Knallgas bacteria such as *C. necator* use the Calvin cycle to assimilate carbon and the enzyme hydrogenase to gain $NAD(P)H + H^+$ from hydrogen as a substrate for the oxidative phosphorylation via the respiratory chain as well as a source for carbon dioxide reduction. Therefore, the overall balance equation for biomass synthesis via PGA including the energy generation from H_2 oxidation at a degree of coupling in the oxidative phosphorylation by the respiratory chain of $P/O = 2$ results in

$$12\ CO_2 + 3\ NH_3 + 56.05\ [H_2] + 15.28\ O_2 \rightarrow 3\ C_4H_8O_2N + 48.55\ H_2O \quad (2)$$

With respect to growth, this corresponds to a carbon conversion efficiency (CCE) of one molecule of carbon (Cmol) incorporated per Cmol supplied and a hydrogen conversion efficiency (HCE) of 0.214 molecules of hydrogen (Hmol) assimilated per Hmol consumed. The synthesis of 2-HIB ($C_4H_8O_3$) as the desired product via the Calvin cycle with PGA and pyruvate as intermediates results in acetyl-CoA (AcCoA) according to

$$4\ CO_2 + 8\ [H_2] + 14\ ATP \rightarrow 2\ AcCoA + 4\ H_2O \quad (3)$$

The ATP required for CO_2 fixation is obtained from hydrogen oxidation via the respiratory chain; accordingly, Equation 3a is extended to

$$4\ CO_2 + 16\ [H_2] + 3.5\ O_2 \rightarrow 2-HIB + 12\ H_2O \quad (4)$$

The CCE is again 1 Cmol/Cmol, whereas the theoretical HCE is 0.25 Hmol/Hmol (Equation 3b). Combining biomass synthesis and product formation to an integral process, the interdependency between both processes defining the final HCE with respect to the product is shown in Figure 1. We took into account two ranges of biomass concentrations (from 0 to 10 g/L and from 10 to 60 g/L) to consider a wide spectrum of variables. Obviously, biomass synthesis is very costly (Equation 2). It is apparent that the overall process approaches a value of 0.2 to 0.25 Hmol/Hmol, when the biomass concentration is below 10 g/L, and the product concentration moves towards 100 g/L (Figure 1). Both the increase in biomass and the reduction of product concentration drastically decrease the HCE.

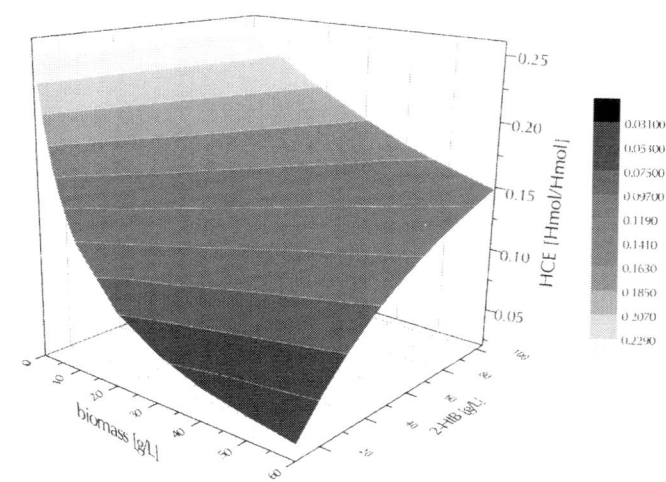

Figure 1: HCE model data. Calculated data for HCE out of biomass concentrations of 0 to 60 g/L and 2-HIB product concentrations of 0 to 100 g/L.

Experimental Data: Growth

The growth characteristics under chemo-litho-autotrophic conditions were examined, thereby displaying a rate of about 0.066/h, which is lower by a factor of about 4 compared to an optimized cultivation regime for the cultivation of C. necator H16 [47,54-56]. Rates reduced by a factor of about 2 are expected, when applying lower temperatures, 22°C in our case compared to 31°C used by former authors. Moreover, the polyhydroxyalkanoate (PHA) synthesis-deficient mutant strain was shown to have a reduced hydrogen oxidation rate compared to the wild type [57]. We used a closed circuit system to recycle the gases in combination with a hollow fiber module as an interface between the gases and the liquid phase for safety reasons due to the explosive character of the gas mixture and due to the necessity of enabling elevated gas transfer conditions under those simplified cultivation conditions. This system has not been further optimized with respect to the transfer rates of the various substrates. Nevertheless, the exponential growth pattern indicates that the substrate supply was not limiting for the biomass concentration applied (Figure 2). Moreover, the results suggest that the cultivation system chosen is in fact adequately efficient in delivering the gaseous substrates for the product synthesis.

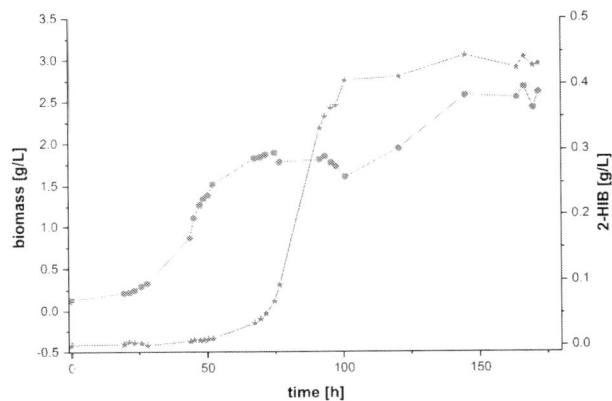

Figure 2: Experimental data for growth and product synthesis of C. necatorH16 PHB−4 (pBBR1MCS-2: HCM) in chemo-litho-autotrophic fermentation. Growth phase (0 to 60 h) and product synthesis phase (60 to 160 h) with biomass (blue circle) and 2-HIB (red star) in g/L.

Experimental Data: Product Formation

We used the strain C. necator H16 PHB⁻4 [47], a PHA-negative mutant, in which the poly-β-hydroxybutyric acid (PHB) synthesis is blocked subsequent to the synthesis of 3-HB-CoA. Introduction of the 2-HIB-CoA mutase from A. tertiaricarbonis allowed this strain to synthesize metabolites up to 3-HB-CoA under conditions of overflow metabolism favoring 2-HIB synthesis through the simultaneous expression of the 2-HIB-CoA mutase. Thereby, an alternative route ensuing 3-HB-CoA is established to finally yield the desired dead-end product, 2-HIB. To confirm the capacity of the chosen system for 2-HIB synthesis, a fructose pre-grown culture was used to inoculate the cultivation apparatus. A gas stream containing 25% to 50% H_2, 15% to 30% CO_2, and 10% to 20% O_2 was supplied as a growth substrate, resulting in the induction of the enzymes required for chemo-litho-autotrophic growth, especially hydrogenases [58,59], and for carbon dioxide fixation [60]. Under these conditions, growth proceeded at a rate of about 0.066/h until the nitrogen source was exhausted, attaining a final biomass concentration of approximately 2.0 g/L (Figure 2). During exponential growth, carbon dioxide was incorporated into the biomass with a CCE of 0.58 Cmol/Cmol. The hydrogen conversion yielded a HCE of 0.0715 Hmol/Hmol. It should be noted that it is not possible to achieve the theoretically maximum value of the HCE of 0.214 Hmol/Hmol due to the required energy (H_2) necessary for maintenance purposes. Larger deviations from the theoretical values might be caused by the synthesis of side products other than biomass, such as pyruvate[57,61], 3-hydroxybutyric acid, acetone, or 2-oxoglutarate [62,63] depending on the culture conditions applied. More detailed analyses, however, were not undertaken at this stage of the investigation.

During growth, 2-HIB was found, but only at low concentrations. After exhaustion of the nitrogen source, however, there was a steep increase in the external 2-HIB concentration (Figure 2). The synthesis rate corresponded to 8.58 mg 2-HIB/[(g bacterial dry mass)·h]. This rate was stable up to a total concentration of approximately 410 mg/L. Subsequently, the product synthesis rate suddenly ceased, which was also observed in repeated experiments. The abrupt shift pointed to a distinct limitation or disturbance, which was not evident in the experimental setup. This would require a further optimization.

As described above, the amount of gases consumed per increment of 2-HIB was used to calculate the yield coefficients. The data were corrected for the unspecific loss of gases determined by running experiments in the absence of biomass. The remaining substrate was incorporated into 2-HIB with a CCE of 0.178 Cmol/Cmol. Hydrogen as a second substrate was converted into this product with a HCE of 0.032 Hmol/Hmol. Noticeably, this is far from the above stated limit values.

The present rates of 2-HIB synthesis with the mutant strain H16 PBH⁻4 (pBBR1MCS-2::HCM) were somewhat lower than those found for the formation of PHB with the wild type strain of *C. necator* H16 under chemo-litho-autotrophic conditions yielding up to 12.8 mg/[(g bacterial dry mass)·h] [64,65]. It is known, however, that the lack of the PHB polymerase (PhaC) activity in this mutant strain is associated with distinctly lower activities of β-ketothiolase (PhaA) and acetoacetyl-CoA reductase (PhaB) [66].

Deficits in Product Synthesis

The deficit in the gain of 2-HIB can likely be explained in part by the putative synthesis of products other than 2-HIB [57,61,63]. Taking into account the reduction of CO_2 by hydrogen to yield the first intermediate of carbon fixation in the Calvin cycle, glyceraldehyde-3-phosphate (GAP, $C_3H_6O_3$, phosphate-free sum formula), the CO_2 available due to the present consumption characteristic will allow for the synthesis of 0.57 mmol GAP/h in the linear phase of 2-HIB formation. This in turn requires 5.16 mmol H_2/h based on

$$3\,CO_2 + 9\,[H_2] + 1.5\,O_2 \rightarrow GAP\,(C_3H_6O_3) + 6\,H_2O \qquad (5)$$

Due to the hydrogen balance, 5.71 mmol H_2 remain available after 2-HIB synthesis which could satisfy the putative product synthesis. Hence, some hydrogen remains additionally available for maintenance purposes which are inevitably necessary in living cells. Taking into account the specific maintenance coefficient determined formerly for *Ralstonia eutropha* (*C. necator*) JMP 134 on fructose of $m_s = 0.09$ mmol/[(g bacterial dry mass)·h] [67] and converting this substrate-based coefficient into an energy (ATP)-based value (P/O = 2) which would be equivalent to $m_e = 2.34$ mmol ATP/[(g bacterial dry mass)·h], the hydrogen remaining after the synthesis of 2-HIB and other putative reduced products would be sufficient to generate 2.8 mmol ATP/[(g bacterial dry mass)·h]. This is in pretty coincidence with former results regarding this species [67].

CONCLUSIONS

The present results can be regarded as a proof of principle demonstrating the feasibility of 2-HIB synthesis under chemo-litho-autotrophic conditions. Since the yield is still far from technological dimensions, the optimization of this process is necessary to improve its stability with the aim to increase the productivity. This would require a prolonged product synthesis, higher rates, and, in particular, better yields.

Under laboratory and chemo-litho-autotrophic conditions, a 2-HIB synthesis rate of 8.58 mg/[(g bacterial dry mass)·h] was achieved, yielding a final concentration of about 0.4 g/L. The efficiency of this production scheme on the basis of sustainable substrates becomes even more obvious when being compared to the derived rates of 2-HIB formation obtained with the same transgenic strain expressing the 2-HIB mutase and being cultivated under aerobic conditions in a controlled fermenter on fructose as the sole substrate. In the latter case, rates of around 5.8 to 7.2 mg 2-HIB/[(g bacterial dry mass)·h] were obtained [45], D Przybylski, unpublished work].

The HCE during product synthesis yielded 0.03 Hmol/Hmol, which is distinctly lower than the theoretical value (0.25 Hmol/Hmol). CO_2 was incorporated into 2-HIB with an efficiency of 0.18 Cmol/Cmol. The discrepancies are not evident at present but are likely to be explained by additional products formed apart from 2-HIB. This follows from the fact that the available amounts of CO_2 and H_2, remaining after 2-HIB synthesis, are sufficient to generate reduced primary products (GAP) in the Calvin cycle in an almost stoichiometric manner (cf. Equation 4).

With respect to the overall yield of the 2-HIB synthesis determined in the present investigation, we extracted a substrate conversion efficiency with a total of 0.103 Cmol/Cmol by taking into account biomass synthesis. This resembles about 63% of the theoretically possible value (0.164 Cmol/Cmol) at the respective biomass (2 g/L) and product concentration (0.4 g/L) according to the treatment of the data as shown in Figure 1. With regard to HCE, the experimental integral value amounted to 0.002 Hmol/Hmol, which is only about 5.5% of the theoretical value with 0.036 Hmol/Hmol. The absolute output of substrate is far too low. This is essentially caused by the low gain of the desired product and an improper ratio between biomass and product concentration.

A question remaining to be solved is the prolonged synthesis of 2-HIB since the present experiment showed a more or less abrupt halt of product synthesis. The reasons for that have to be thoroughly examined in order to be possibly eliminated in future experiments.

The present investigation was performed under laboratory conditions in a 0.6-L dimension and at a low biomass concentration of around 2 g/L. Upscaling will illustrate the potential of such a biobased process. Based on the specific rate of 8.58 mg 2-HIB/[(g bacterial dry mass)·h] found in the present investigation, a process extrapolated to the cubic meter dimension would therefore result in the synthesis of approximately 200 g 2-HIB/(m^3 d) by applying 1 kg of biomass. Using 10 kg of biomass and a 10-m^3 scale, the output will be 20 kg/day, which corresponds to a production on a semi-technical scale. In general, the chemical industry operates reactors with a size of 1,000 m^3 and larger. Due to the obvious reasons, more and more processes will and already do involve gases and thus require experience in handling explosive mixtures. They will not likely involve a membrane technology, as has been used here for safety reasons. Consequently, amounts of tons per day are imaginable without relying on unrealistic assumptions. Higher biomass concentrations will have an even higher impact on the productivity of such a process. In this case, the efficiency of the conversion of the substrates to the final product has to be considered (see Figure 1). An increase in biomass concentration will consequently result in a diminished efficiency of the product synthesis. An optimization at this stage will include considerations about rate versus yield, subsequently leading to decisions based on economic figures.

Another important factor not to be neglected in this context is the usability/durability of the catalyst biomass. The present case assumes a discontinuous production regime since organic acids, as the envisaged product 2-HIB, are in general toxic to microorganisms at higher concentrations [68-72], e.g., acetic acid being inhibitive above concentrations of 6 g/L [73], which is well below the desired product concentration range. However, no thorough investigations with respect to matters of 2-HIB product inhibition have been undertaken at this stage of investigation. But as acid toxicity will have an impact and therefore has to be considered, continuous extraction offers the possibility to reduce the current acid concentration in the production broth with the effect of maintaining the activity of the cells and thus extending the production time. Moreover, a continuous process with intermittent periods of growth to regenerate the catalyst biomass should be considered. Such measures and their effects, however, require detailed investigations which are outside the scope of the present investigation.

The yearly production of Plexiglas® amounts to 3 million tons and is based exclusively on fossil carbon sources. However, it is not imaginable that a process as described here will substitute the established processes in the near future, but the actual constellations contribute to a turnaround in the favor of alternative processes relying on gases. Production and storage of hydrogen on the basis of electricity generated by solar techniques and wind energy is state of the art and will increase in its dimension [35,36,74]. Carbon dioxide, on the other hand, is an unavoidable result of energy production from fossil carbon sources. As the actual discussions address the question of how to get rid of this climate change driver, the removal of this compound from exhaust gases in energy plants and its deposition are currently argued for, and legislation will find ways to implement such solutions. Above all, those factors emphasize even more the necessity of a turn in thinking regarding the handling of our resources while supporting new ideas and developments to move in the direction suggested in this investigation. Moreover, gases derived from biomass, i.e., synthesis gas comprised of hydrogen and carbon monoxide as a result of pyrolysis [4,75,76], are yet another source for product syntheses as the one described here.

Adequate pathways are found in diverse microorganisms, such as in anaerobic clostridia. Forthcoming models following the idea of sustainable product synthesis could be directed towards methane as a potential substrate [4] as well, which may be derived from biogas plants or received as a conversion product out of synthesis gas and methanol as its oxidized derivative. The basic metabolic potentials to convert those substrates are available in the respective microorganisms. The decision regarding the kind of substrate to be used for an envisaged product synthesis depends on technological and physiological properties. The final decision then depends on the degree of the required refinement of the educts, which in turn is a question of the effectiveness of substrate conversion and the price of the final product [67].

AUTHORS' CONTRIBUTIONS

DP conceived the study, participated in its design, and drafted the manuscript. TR and HH participated in the concept and drafted the manuscript. RHM participated in the experimental design, contributed to the model, and drafted the manuscript. All authors have read and approved the final manuscript.

DP is a doctoral student. She is dealing with biotechnological issues on the application and optimization of properties of the conversion of microbial substrates into distinct products by applying enzymatic, physiological, fermentative, and thermodynamic techniques.

ACKNOWLEDGMENTS

This investigation was financed by the Fachagentur Nachwachsende Rohstoffe e.V. (FNR)/Bundesministerium für Ernährung, Landwirtschaft und Verbraucherschutz (BMELV), grant FKZ 22027805. The authors are furthermore grateful to the Evonik Industries GmbH for the support. Additionally, the work of DP was carried out under the mentoring of the HIGRADE Graduate School.

REFERENCES

1.   Zinoviev S, Mueller-Langer F, Das P, Bertero N, Fornasiero P, Kaltschmitt M, Centi G, Miertus S (2010) Next-generation biofuels: survey of emerging technologies and sustainability issues. Chem Sus Chem 3(10):1106-1133

2.   Willke T, Vorlop KD (2004) Industrial bioconversion of renewable resources as an alternative to conventional chemistry. Appl Microbiol Biotechnol 66(2):131-142

3.   Nordhoff S, Hocker H, Gebhardt H (2007) Renewable resources in the chemical industry – breaking away from oil? Biotechnol J 2(12):1505-1513

4.   Muffler K, Ulber R (2008) Use of renewable raw materials in the chemical industry – beyond sugar and starch. Chem Eng Technol 31(5):638-646

5.   Hatti-Kaul R, Tornvall U, Gustafsson L, Borjesson P (2007) Industrial biotechnology for the production of bio-based chemicals – a cradle-to-grave perspective. Trends Biotechnol 25(3):119-124

6.   Roes AL, Patel MK (2007) Life cycle risks for human health: a comparison of petroleum versus bio-based production of five bulk organic chemicals. Risk Anal 27(5):1311-1321

7.   Dornburg V, Hermann BG, Patel MK (2008) Scenario projections for future market potentials of biobased bulk chemicals. Environ Sci Technol 42(7):2261-2267

8.   Hahn-Haegerdal B, Galbe M, Gorwa-Grauslund MF, Lidén G, Zacchi G (2006) Bio-ethanol – the fuel of tomorrow from the residues of today. Trends Biotechnol 24(12):549-556

9.   Otero JM, Panagiotou G, Olsson L (2007) Fueling industrial biotechnology growth with bioethanol. Adv Biochem Eng Biotechnol 108:1-40

10.   Mussatto SI, Dragone G, Guimaraes PM, Silva JP, Carneiro LM, Roberto IC, Vicente A, Domingues L, Teixeira JA (2010)

Technological trends, global market, and challenges of bio-ethanol production. Biotechnol Adv 28(6):817-830

11. Singh A, Pant D, Korres NE, Nizami AS, Prasad S, Murphy JD (2010) Key issues in life cycle assessment of ethanol production from lignocellulosic biomass: challenges and perspectives. Bioresource Technol 101(13):5003-5012

12. Kohse-Hoinghaus K, Osswald P, Cool TA, Kasper T, Hansen N, Qi F, Westbrook CK, Westmoreland PR (2010) Biofuel combustion chemistry: from ethanol to biodiesel. Angew Chem Int Ed Engl 49(21):3572-3597

13. Dürre P (2008) Fermentative butanol production: bulk chemical and biofuel. Ann N Y Acad Sci 1125:353-362

14. Dong H, Tao W, Dai Z, Yang L, Gong F, Zhang Y, Li Y (2011) Biobutanol. Adv Biochem Eng Biotechnol.

15. Werpy T, Petersen B (2004) Top value added chemicals from biomass 1. , . Available via http://www.eere.energy.gov/biomass/pdfs/pnnl-16983.pdf. Accessed 27 June 2012

16. Busch R, Hirth T, Liese A, Nordhoff S, Puls J, Pulz O, Sell D, Syldatk C, Ulber R (2006) The utilization of renewable resources in German industrial production. Biotechnol J 1(7–8):770-776

17. Marner WD (2009) Practical application of synthetic biology principles. Biotechnol J 4(10):1406-1419

18. Tang WL, Zhao H (2009) Industrial biotechnology: tools and applications. Biotechnol J 4(12):1725-1739

19. Zeng AP, Biebl H (2002) Bulk chemicals from biotechnology: the case of 1,3-propanediol production and the new trends. Adv Biochem Eng Biotechnol 74:239-259

20. McKinlay JB, Vieille C, Zeikus JG (2007) Prospects for a bio-based succinate industry. Appl Microbiol Biotechnol 76(4):727-740

21. Anastassiadis S, Morgunov IG (2007) Gluconic acid production. Recent Pat Biotechnol 1(2):167-180

22. Steudel A, Miethe D, Babel W (1980) Bacterium MB 58, a methylotrophic "acetic acid bacterium". Z Allg Mikrobiol 20(10):663-672

23. Ramachandran S, Fontanille P, Pandey A, Larroche C (2006) Gluconic acid: a review: properties, applications and microbial production. Food Technol Biotechnol 44(2):185-195

24. Anastassiadis S, Morgunov IG, Kamzolova SV, Finogenova TV (2008) Citric acid production patent review. Recent Pat Biotechnol 2(2):107-123

25. Müller RH, Rohwerder T, Harms H (2007) Carbon conversion efficiency and limits of productive bacterial degradation of methyl tert-butyl ether and related compounds. Appl Environ Microbiol 73(6):1783-1791

26. Bai FW, Anderson WA, Moo-Young M (2008) Ethanol fermentation technologies from sugar and starch feedstocks. Biotechnol Adv 26(1):89-105

27. Naik SN, Goud VV, Rout PK, Dalai AK (2010) Production of first and second generation biofuels: a comprehensive review. Renew Sust Energ Rev 14(2):578-597

28. Demain AL (2009) Biosolutions to the energy problem. J Ind Microbiol Biotechnol 36(3):319-332

29. Kanowski J, Catterall CP (2010) Carbon stocks in above-ground biomass of monoculture plantations, mixed species plantations and environmental restoration plantings in north-east Australia. Ecol Manag Restor 11(2):119-126

30. Lenk F, Broring S, Herzog P, Leker J (2007) On the usage of agricultural raw materials – energy or food? An assessment from an economics perspective. Biotechnol J 2(12):1497-1504

31. Thrän D, Kaltschmitt M (2007) Competition – supporting or preventing an increased use of bioenergy? Biotechnol J 2(12):1514-1524

32. Carere C, Sparling R, Cicek N, Levin D (2008) Third generation biofuels via direct cellulose fermentation. Int J Mol Sci 9(7):1342-1360

33. Demirbas FM (2009) Biorefineries for biofuel upgrading: a critical review. Appl Energy 86 Supplement 1(0):S151-S161

34. Hermann BG, Blok K, Patel MK (2007) Producing bio-based bulk chemicals using industrial biotechnology saves energy and combats climate change. Environ Sci Technol 41(22):7915-7921

35. Bak T, Nowotny J, Rekas M, Sorrell CC (2002) Photo-electrochemical hydrogen generation from water using solar energy. Materials-related aspects. Int J Hydrogen Energy 27(10):991-1022

36. Sherif SA, Barbir F, Veziroglu TN (2005) Wind energy and the hydrogen economy – review of the technology. Sol Energy 78(5):647-660

37. Peters M, Köhler B, Kuckshinrichs W, Leitner W, Markewitz P, Müller TE (2011) Chemical technologies for exploiting and recycling carbon dioxide into the value chain. Chem Sus Chem 4(9):1216-1240

38. Cokoja M, Bruckmeier C, Rieger B, Herrmann WA, Kühn FE (2011) Transformation of carbon dioxide with homogeneous transition-metal catalysts: a molecular solution to a global challenge? Angew Chem Int Ed Engl 50(37):8510-8537

39. Pelzer S (2012) Maßgeschneiderte Mikroorganismen. Biol Unserer Zeit 42(2):98-106

40. Hermann BG, Patel M (2007) Today's and tomorrow's bio-based bulk chemicals from white biotechnology: a techno-economic analysis. Appl Biochem Biotechnol 136(3):361-388

41. Rohwerder T, Breuer U, Benndorf D, Lechner U, Müller RH (2006) The alkyl tert-butyl ether intermediate 2-hydroxyisobutyrate is degraded via a novel cobalamin-dependent mutase pathway. Appl Environ Microbiol 72(6):4128-4135

42. Rohwerder T, Müller RH (2007) New bacterial cobalamin-dependent CoA-carbonyl mutases involved in degradation

pathways. In: Elliot CM (ed) Vitamin B: new research, Nova Science Publishers, New York. pp 81-98

43. Müller RH, Rohwerder T (2007) Method for the enzymatic production of 2-hydroxy-2-methyl carboxylic acids. http://patentscope.wipo.int/search/en/WO2007110394 *webcite*. Accessed 27 June 2012

44. Yaneva N, Schuster J, Schäfer F, Lede V, Przybylski D, Paproth T, Harms H, Müller RH, Rohwerder T (2012) A bacterial acyl-CoA mutase specifically catalyzes coenzyme B12-dependent isomerization of 2-hydroxyisobutyryl-CoA and (S)-3-hydroxybutyryl-CoA. J Biol Chem.

45. Höfel T, Wittmann E, Reinecke L, Weuster-Botz D (2010) Reaction engineering studies for the production of 2-hydroxyisobutyric acid with recombinant Cupriavidus necator H 16. Appl Microbiol Biotechnol 88(2):477-484

46. Schlegel HG (1954) The role of molecular hydrogen in the metabolism of microorganisms. Arch Mikrobiol 20(3):293-322

47. Schlegel HG, Lafferty R, Krauss I (1970) The isolation of mutants not accumulating poly-beta-hydroxybutyric acid. Arch Mikrobiol 71(3):283-294

48. Reinecke L, Schaffer S, Marx A, Pötter M, Haas T (2009) Recombinant cell producing 2-hydroxyisobutyric acid. http://patentscope.wipo.int/search/en/WO2009156214 *webcite*. Accessed 27 June 2012

49. Kovach ME, Elzer PH, Hill DS, Robertson GT, Farris MA, Roop RM, Peterson KM (1995) Four new derivatives of the broad-host-range cloning vector pBBR1MCS, carrying different antibiotic-resistance cassettes. Gene 166(1):175-176

50. Schlegel HG, Kaltwasser H, Gottschalk G (1961) Ein Submersverfahren zur Kultur wasserstoffoxydierender Bakterien: Wachstumsphysiologische Untersuchungen. Arch Microbiol 38(3):209-222

51. van Dijken JP, Harder W (1975) Growth yields of microorganisms on methanol and methane. A theoretical study. Biotechnol Bioeng 17(1):15-30

52. Babel W, Müller RH (1985) Correlation between cell composition and carbon conversion efficiency in microbial growth: a theoretical study. Appl Microbiol Biotechnol 22(3):201-207

53. Stouthamer AH (1973) A theoretical study on the amount of ATP required for synthesis of microbial cell material. A Van Leeuw J Microb 39(3):545-565

54. Repaske R, Mayer R (1976) Dense autotrophic cultures of Alcaligenes eutrophus. Appl Environ Microbiol 32(4):592-597

55. Ishizaki A, Tanaka K (1991) Production of poly- -hydroxybutyric acid from carbon dioxide by Alcaligenes eutrophus ATCC 17697 T. J Ferment Bioeng 71(4):254-257

56. Ishizaki A, Tanaka K (1990) Batch culture of Alcaligenes eutrophus ATCC 17697 T using recycled gas closed circuit culture system. J Ferment Bioeng 69(3):170-174

57. Steinbüchel A, Schlegel HG (1989) Excretion of pyruvate by mutants of Alcaligenes eutrophus, which are impaired in the accumulation of poly(-hydrcxybutyric acid) (PHB), under conditions permitting synthesis of PHB. Appl Microbiol Biotechnol 31(2):168-175

58. Lenz O, Schwartz E, Dernecde J, Eitinger M, Friedrich B (1994) The Alcaligenes eutrophus H16 hoxX gene participates in hydrogenase regulation. J Bacteriol 176(14):4385-4393

59. Burgdorf T, Lenz O, Buhrke T. van der Linden E, Jones AK, Albracht SP, Friedrich B (2005) [NiFe]-hydrogenases of Ralstonia eutropha H16: modular enzymes for oxygen-tolerant biological hydrogen oxidation. J Mol Microbiol Biotechnol 10(2–4):181-196

60. Husemann M, Klintworth R, Büttcher V, Salnikow J, Weissenborn C, Bowien B (1988) Chromosomally and plasmid-encoded gene clusters for CO2 fixation (cfx genes) in Alcaligenes eutrophus. Mol Gen Genet MGG 214(1):112-120

61. Vollbrecht D, Nawawy MA, Schlegel HG (1978) Excretion of metabolites by hydrogen bacteria I. Autotrophic and heterotrophic fermentations. Appl Microbiol Biotechnol 6(2):145-155

62. Vollbrecht D, Schlegel HG (1978) Excretion of metabolites by hydrogen bacteria II. Influences of aeration, pH, temperature, and age of cells. Appl Microbiol Biotechnol 6(2):157-166

63. Vollbrecht D, Schlegel HG, Stoschek G, Janczikowski A (1979) Excretion of metabolites by hydrogen bacteria. Appl Microbiol Biotechnol 7(3):267-276

64. Tanaka K, Ishizaki A, Kanamaru T, Kawano T (1995) Production of poly(d-3-hydroxybutyrate) from CO(2), H(2), and O(2) by high cell density autotrophic cultivation of Alcaligenes eutrophus. Biotechnol Bioeng 45(3):268-275

65. Sonnleitner B, Heinzle E, Braunegg G, Lafferty RM (1979) Formal kinetics of poly- -hydroxybutyric acid (PHB) production in Alcaligenes eutrophus H 16 and Mycoplana rubra R 14 with respect to the dissolved oxygen tension in ammonium-limited batch cultures. Appl Microbiol Biotechnol 7(1):1-10

66. Schubert P, Steinbuchel A, Schlegel HG (1988) Cloning of the Alcaligenes eutrophus genes for synthesis of poly-beta-hydroxybutyric acid (PHB) and synthesis of PHB in Escherichia coli. J Bacteriol 170(12):5837-5847

67. Müller RH, Babel W (1996) Measurement of growth at very low rates ($\mu \geq 0$), an approach to study the energy requirement for the survival of Alcaligenes eutrophus JMP 134. Appl Environ Microbiol 62(1):147-151

68. Lee SE, Li QX, Yu J (2009) Diverse protein regulations on PHA formation in Ralstonia eutropha on short chain organic acids. Int J Biol Sci 5(3):215-225

69. Russell JB (1992) Another explanation for the toxicity of fermentation acids at low pH: anion accumulation versus uncoupling. J Appl Microbiol 73(5):363-370

70. Chung YJ, Cha HJ, Yeo JS, Yoo YJ (1997) Production of poly(3-hydroxybutyric-co-3-hydroxyvaleric)acid using propionic acid by pH regulation. J Ferment Bioeng 83(5):492-495

71. Wang J, Yu J (2000) Kinetic analysis on inhibited growth and poly(3-hydroxybutyrate) formation of Alcaligenes eutrophus on acetate under nutrient-rich conditions. Process Biochem 36(3):201-207

72. Du G, Si Y, Yu J (2001) Inhibitory effect of medium-chain-length fatty acids on synthesis of polyhydroxyalkanoates from volatile fatty acids by Ralstonia eutropha. Biotechnol Lett 23(19):1613-1617

73. Yu J, Wang J (2001) Metabolic flux modeling of detoxification of acetic acid by Ralstonia eutropha at slightly alkaline pH levels. Biotechnol Bioeng 73(6):458-464

74. Greiner CJ, KorpÅs M, Holen AT (2007) A Norwegian case study on the production of hydrogen from wind power. Int J Hydrogen Energy 32(10–11):1500-1507

75. Phillips J, Clausen E, Gaddy J (1994) Synthesis gas as substrate for the biological production of fuels and chemicals. Appl Biochem Biotechnol 45–46(1):145-157

76. Digman B, Joo HS, Kim DS (2009) Recent progress in gasification/pyrolysis technologies for biomass conversion to energy. Environ Prog & Sustainable Energy 28(1):47-51

Focus on Potential Environmental Issues on Plastic World towards a Sustainable Plastic Recycling in Developing Countries

Onwughara Innocent Nkwachukwu[1, 3], Chukwu Henry Chima[2], Alaekwe Obiora Ikenna[3], and Lackson Albert[3, 4]

[1]Reliable Research Laboratory Service, D30 Orji Kalu Housing Estate, Umuahia, Abia State, Nigeria

[2]Department of Chemistry, Abia State Polytechnic, Aba, Abia State, Nigeria

[3]Department of Pure and Industrial Chemistry, Nnamdi Azikiwe University Awka, P.M.B. 5052, Awka, Anambra State, Nigeria

[4]Yagai Academy, P.O. Box 1180, Jalingo, Taraba State, Nigeria

ABSTRACT

Due to the tremendous growth of plastics in the world, it has brought about environmental concerns for over the past two or three decades. Most of these plastics, due to poor management, are currently disposed of in unauthorized dumping sites or burned uncontrollably in the fields. The paper outlines environmental concerns of so many applications of plastics. The most important mechanisms of degradation of plastics, environmental impacts and recommendations for sustainable development are fully discoursed, with recycling option being overviewed as the route under most intense development at this time because of its broad public appeal and obvious environmental advantages.

REVIEW

Introduction

Plastics are organic polymeric materials consisting of giant organic molecules. Plastic materials can be formed into shapes by one of a variety of processes, such as extrusion, moulding, casting or spinning. Modern plastics (or polymers) possess a number of extremely desirable characteristics: high strength-to-weight ratio, excellent thermal properties, electrical insulation, and resistance to acids, alkalis and solvents, to name but a few. Some have unique electrical insulating properties, such as their strength, stress resistance, flexibility and durability, which make them important materials for use in electronics. These polymers are made of a series of repeating units known as monomers. The structure and degree of polymerisation of a given polymer determine its characteristics. Linear polymers (a single linear chain of monomers) and branched polymers (linear with side chains) are thermoplastic; they soften when heated. Cross-linked polymers (polymers with bond formed between polymer chains, either between different chains or between different parts of the same chain.) are thermosetting, that is, they harden when heated.

Development of synthetic polymers, used to make plastics such as polyethylene, polypropylenes, polyesters and polyamides (including nylon), has revolutionized the types of containers for products, the types of materials for packaging and the materials used for carry bags. However, most of these polymers are not biodegradable and, once used and discarded, become major waste management challenges [1]. However, plastic waste can also impose negative externalities such as greenhouse gas emissions or ecological damage. It is usually non-biodegradable and therefore can remain as waste in the environment for a very long time; it may pose risks to human health and the environment. In some cases, it can be difficult to reuse and/or recycle. There is a mounting body of evidence which indicates that substantial quantities of plastic waste are now polluting marine and other habitats [2]. The widespread presence of these materials has resulted in numerous accounts of wildlife becoming entangled in plastic, leading to injury or impaired movement and, in some cases, resulting in death. Concerns have been raised regarding the effects of plastic ingestion as there is some evidence to indicate that toxic chemicals from plastics can accumulate in living organisms and throughout nutrient chains. There is also some public health concerns arising from the use of plastics treated with chemicals [2].

As with most materials, global plastic production is estimated to decrease from 245 million tonnes (Mt) in 2008 to around 230 million tonnes in 2009 as a result of the economic crisis.Over the past 50 years, however, there has been a very steep rise in plastic production, especially in Asia (Figure 1). The European Union (EU) accounts for around 25% of world production. The total consumption of plastics in Western Europe was approximately 39.7 million tonnes in 2003, which means about 98 kg/person, and the amount has been increasing [3]. China produces more plastic than any other country, at 15% of global production. Germany produces the greatest amount of any EU country, accounting for about 8% of global production as shown in Figure 2[4]. Societies are increasingly reliant on plastics, which are already a ubiquitous part of everyday life. As the development of new materials is ongoing, limiting their detrimental effects poses new challenges for policy makers. Regulatory instruments designed to mitigate the effects of plastics on human health and the environment must evolve in line with trends in production, use and disposal [2].

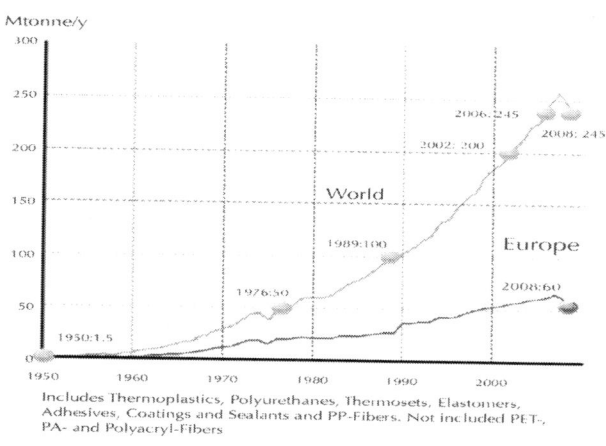

Figure 1: World plastic production, 1950 to 2008 (Mt) (adapted from [[5]]).

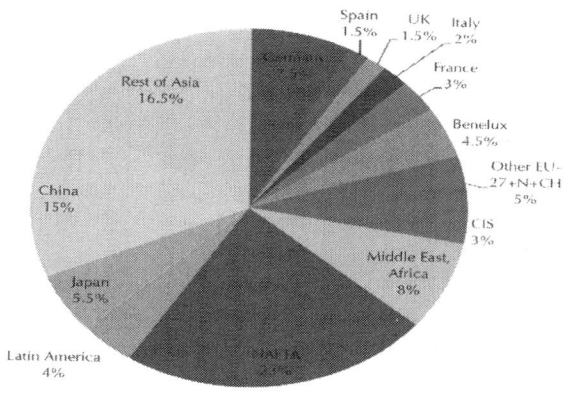

Figure 2: Distribution of world plastic production (adapted from [[4]]).

Polyethylene has the highest share of production of any polymer type, while four sectors: polyethylene terephthalate (PET), which accounts for 20% of thermoplastic resin capacity, followed by polypropylene (PP), which accounts for 18%, polyvinyl chloride (PVC) and polystyrene/expanded polystyrene (PS/EPS), represent 72% of plastic demand: packaging, construction, automotive and electrical and electronic equipment as shown in Figure 3. The rest includes sectors such as household, furniture, agriculture and medical devices [4]. Plastic packaging accounts for the largest share of plastic production in the world level. About 50% of plastic is used for single-use disposable applications, such as packaging, agricultural films and disposable consumer items [6]. Plastics were the second largest component in waste from electrical and electronic equipment (WEEE), and approximately 30% of the mass electronic scrap consists of plastics [7-9]. Plastics consume approximately 8% of world oil production: 4% as raw material for plastics and 3% to 4% as energy for manufacture [1, 6].

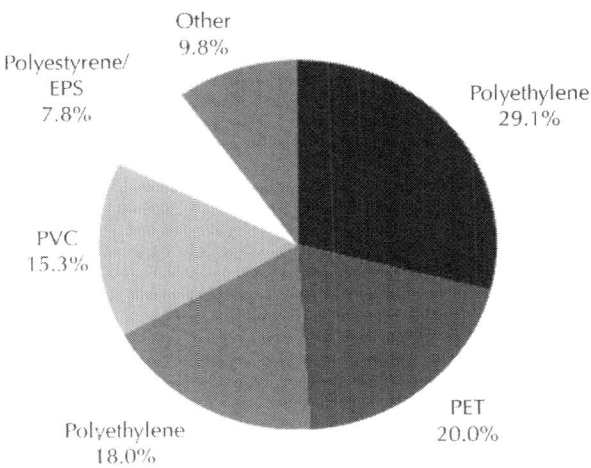

Figure 3: World thermoplastic resin capacity, 2008 (adapted from [[10]]).

The plastic industry is in constant development, with technology evolving in response to the ever-changing demand. Some trends that emerge clearly are continued innovation and improvements such as weight reduction of individual items, increasing use of plastics (and bioplastics) in vehicle manufacturing, a shift in primary plastic production to transition and emerging economies and continued growth in the market share of bioplastics (despite some sorting and price barriers).

Bioplastics make up only 0.1% to 0.2% of total EU plastics [11]. It is estimated that plastics save 600 to 1,300 million tonnes of CO_2 through the replacement of less efficient materials, fuel savings in transport, contribution to insulation, prevention of food losses and use in wind power rotors and solar panels [12]. In 2000, the consumption of polymers for plastic applications in Western Europe was 36,769,000 tonnes, an increase of 3.4% from 1999 [13]. Of the generated municipal solid waste (MSW) in Thailand, 14% were plastics [14]. According to Onwughara [15], the percentage components of plastics and nylon of different categories of solid generated in Umuahia, capital of Abia State, Nigeria were 1.5% and 10.2%, respectively. Of the generated wastes in Kathmandu Valley in Nepal, 22.65% were plastics [16].

The chemicals produced known as dioxins and furan from plastic, especially incinerating plastics, have been implicated in birth defects and several kinds of cancer. The slag and fly ash were found to be environmentally beneficial in cement production and for off-gases for power production [17]. Thermoplastics make up 80% of the plastics produced today [17]. Examples of thermoplastics include high-density polyethylene (HDPE) used in piping, automotive fuel tanks, bottles and toys; low-density polyethylene (LDPE) used in plastic bags, cling film and flexible containers; PET used in bottles, carpets and food packaging; PP used in food containers, battery cases, bottle crates, automotive parts and fibres; PS used in dairy product containers, tape cassettes, cups and plates; and PVC used in window frames, flooring, bottles, packaging film, cable insulation, credit cards and medical products.

There are hundreds of types of thermoplastic polymer, and new variations are regularly being developed. In developing countries, the number of plastics in common use, however, tends to be much lower. Thermosets make up the remaining 20% of plastics produced.

They are hardened by curing and cannot be re-melted or re-moulded and are therefore difficult to recycle. They are sometimes ground and used as a filler material. They include polyurethane (PU) - coatings, finishes, gears, diaphragms, cushions, mattresses and car seats; epoxy - adhesives, sports equipment and electrical and automotive equipment; and phenolics - ovens, handles for cutlery, automotive parts and circuit boards. The global demand for plastic composites has grown significantly over the past few years (Figure 4).

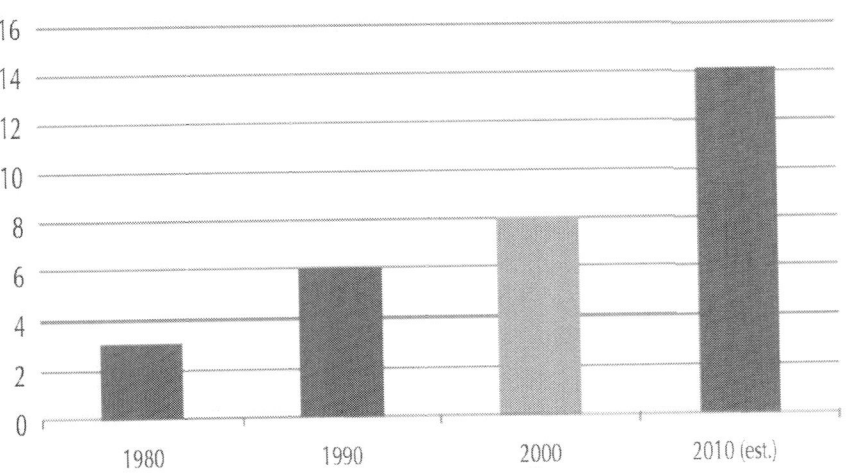

Figure 4: Global demand in the composite industry (Mt) (adapted from [[18]]).

Nowadays, the raw materials for plastics come mainly from petrochemicals, although originally plastics were derived from cellulose, the basic material of all plant life. The materials used in electronics have several important characteristics. In PC monitors and televisions (TVs), acrylonitrile butadiene styrene (ABS) and high-impact polystyrene (HIPS) are used for cathode ray tube protection. Also, polyphenylene oxide (PPO) has good properties such as high temperature resistance, rigidity, impact strength and creep resistance.

Table 1 shows the summary of typical resins used in different electrical and electronics equipment[7], and Table 2 shows the weight percentage of manufactured plastic from organic compounds[7,15]. Polymer types used in various construction applications are described in Table 3.

Table 1: Resins used in electronic products

EEE	Resins
Computers	ABS, HIPS, PPO, PPE, PVC, PC/ABS
TVs	HIPS, PC, ABS, PPE, PVC
Miscellaneous	HIPS, PVC, ABS, PC/ABS, PPE, PC

ABS acrylonitrile butadiene styrene, HIPS high-impact polystyrene, PPO polyphenylene oxide, PPE polyphenylene ether, PVC polyvinylchloride, PC polycarbonate.Miscellaneous: fax, telephone, refrigerator etc.

Nkwachukwu et al.

Nkwachukwu et al. International Journal of Industrial Chemistry 2013 4:34 doi: 10.1186/2228-5547-4-34.

Table 2: Manufactured plastic from organic compounds

Manufactured plastic	Weight percentage (wt.%)
HIPS	59
ABS	20
PPO	16
PP or PE	2
Others	3

HIPS high-impact polystyrene, ABS acrylonitrile butadiene styrene, PPO polyphenylene oxide, PP polypropylene, PE polyethylene.

Nkwachukwu et al.

Nkwachukwu et al. International Journal of Industrial Chemistry 2013 4:34 doi: 10.1186/2228-5547-4-34.

Table 3: Main polymers used for applications

Application	Most common polymers used
Pipes and ducts	PVC, PP, HDPE, LDPE, ABS
Insulation	PU, EPS, XPS
Window profiles	PVC
Other profiles	
Floor and wall coverings	
Lining	PE, PVC
Fitted furniture	PS, PMMA, PC, POM, PA, UP, amino

PVC polyvinylchloride, PP polypropylene, HDPE high-density polyethylene, LDPE low-density polyethylene, ABS acrylonitrile butadiene styrene, PU polyurethane, EPS expanded polystyrene, XPS extruded polystyrene.

Nkwachukwu et al.

Nkwachukwu et al. International Journal of Industrial Chemistry 2013 4:34 doi: 10.1186/2228-5547-4-34.

A more recent projection (Figure 5) shows slightly slower growth to just over 1.4 Mt in 2013, but the trend is still strongly positive. The SRI study projects total consumption of biodegradable polymers worldwide at an average annual growth rate of 13% from 2009 to 2014 [19].

According to Kurudufu [21], it is estimated that 100 million tonnes of plastics are produced each year. The average European throws away 36 kg of plastics each year. Four percent of oil consumption in Europe is used for the manufacture of plastic products. Some plastic waste sacks are made from 64% recycled plastic. Plastic packaging totals 42% of the total consumption, and very little of this is recycled.

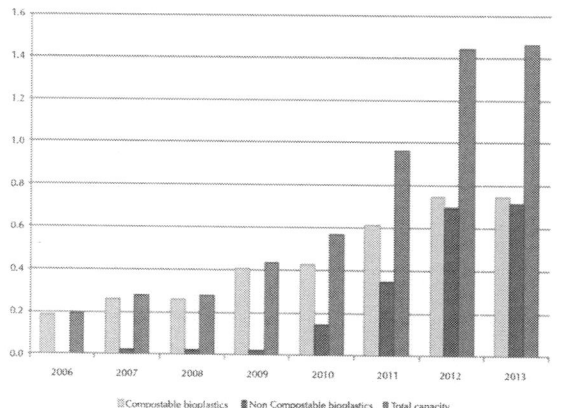

Figure 5: Global production capacity of bioplastics (Mt) (adapted from [[20]]).

In 2008, total generation of post-consumer plastic waste in EU-27, Norway and Switzerland was 24.9 Mt. Packaging is by far the largest contributor to plastic waste at 63%. Average EU-27 per capita generation of plastic packaging waste was 30.6 kg in 2007 [4]. There are lots of different plastics, and they will give off lots of different vapours when they decompose. Figure 6 shows various areas where plastics are used [22].

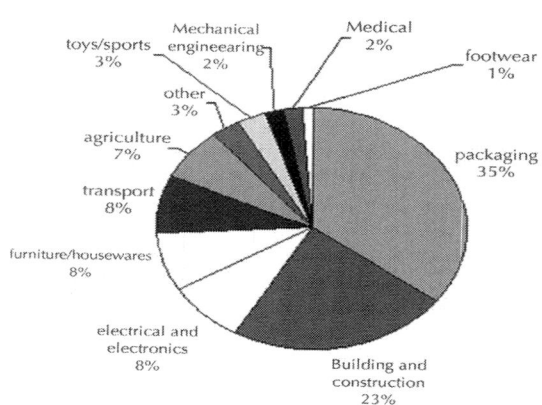

Figure 6: Utilization of plastic in various fields.

It could be just a simple hydrocarbon or it could contain cyanides, polychlorinated biphenyls (PCBs) or lots of other substances. Without knowing what the plastic was (including what additives might have been incorporated), it would be difficult to know what likely volatiles it would create; volatiles given off from plastics in house fires are a major cause of death. Halogenated plastics, those that are made from chlorine or fluorine, are problematic. This work will review environmental issue ascertained from the development of plastics.

Sources of Waste Plastics

Industrial waste (or primary waste) can often be obtained from the large plastic processing, manufacturing and packaging industries. Rejected or waste material usually has good characteristics for recycling and thus will be clean. Although the quantity of material available is sometimes small, the quantities tend to be growing as consumption, and therefore production, increases. Commercial waste is often available from workshops, craftsmen, shops, supermarkets and wholesalers. A lot of the plastics available from these sources will be PE, often contaminated. Agricultural waste can be obtained from farms and nursery gardens outside the urban areas. This is usually in the form of packaging (plastic containers or sheets) or construction materials (irrigation or hosepipes). Municipal waste can be collected from residential areas (domestic or household waste), streets, parks, collection depots and waste dumps. In Asian cities, this type of waste is common and can either be collected from the streets or from households by arrangement with the householders [21].

Plastics' End-of-life

Several end-of-life options exist to deal with plastic waste, including recycling, disposal and incineration with or without energy recovery. The plastic recycling rate was 21.3% in 2008, helping to drive total recovery (energy recovery and recycling) to 51.3%. The highest rate of recycling is seen in Germany at 34% [4]. As plastic packaging has the longest established system for the recovery and recycling of plastic waste, it is natural that its recycling rates are higher than those of other streams. It is followed by agricultural waste plastic, which, although not under direct legislative obligation to increase recovery, is subject to economic incentives linked to the availability of homogenous materials. Although WEEE and construction plastic waste sources have relatively low rates of recycling overall, the rate of energy recovery is relatively high. Overall, total recovery is the highest for plastic packaging at 59.8% and the lowest for ELV plastics at 19.2% [4].

The final stage in the life cycle of plastics is disposal. In India, there are three common ways of getting rid of plastics: by dumping them in landfills, by burning them in incinerators or by littering them. In the case of littering, plastic wastes fail to reach landfills or incinerators. It is the improper way of disposing plastics and is identified as the cause of manifold ecological problems. Incineration is a process in which plastic and other wastes are burnt, and the energy produced, as a result, is tapped. In combination with halogens in the plastic fraction, they can form volatile metal halides, but they also have a catalytic effect on the formation of dioxins and furans [9]. WEEE should not be combined with unsorted municipal waste destined for landfills or open burning of garbage because electronic waste can contain more than 100 different substances (toxic chemicals), many of which are toxic such as lead, mercury, hexavalent chromium, selenium, cadmium and arsenic [23]. Additional harmful substances in WEEE can include arsenic, PCBs, chlorofluorocarbons (CFCs) and hydrochlorofluorocarbons (HCFCs) and nickel. Some of these toxic chemicals, even when present in small amounts, can be potent pollutants and contribute to toxic landfill leachate and vapours, such as vaporization of metallic and dimethylene mercury [15].

During burning of WEEE, toxic chemicals such as dioxins and furans may be release to the environment; furthermore, runoff water carries leachate from acidic ash into the sea affecting the aquatic life. Also, the ash leached into the soil which causes groundwater contamination. The municipal solid wastes in Nigeria contain all sources of unsorted wastes, such as commercial refuse, construction and demolition debris, garbage, electronic wastes etc., which are dumped indiscriminately on roadsides and any available open pits irrespective of the health implication on people [15]. Most plastics (thermosets) are from electronic wastes [15]. With the rapid improvements in the electronic industry, electronic waste (e.waste), including all obsolete electronic products, has become the fastest growing component in the solid waste stream. This phenomenon has been a source of hazardous wastes such as PCs and TVs, which contain heavy metals and organic compounds that pose risk to the environment if not properly managed. Balakrishnan shows that 19% of plastic are found in WEEE [8].

More than 20,000 plastic bottles are needed to obtain 1 tonne of plastic [24]. The durability of the most widely used polyethylene plastic films used for protected cultivation varies from a minimum of one cultivating season to a maximum of 2 to 3 years, and at the end of their useful life, they are classified as waste. Most of this waste is currently disposed of in unauthorized dumping sites or burned uncontrollably in the fields. Management of the huge quantities of waste produced in this way represents a problem with great environmental implications. In order to minimize the environmental impact of this plastic waste stream, it is desirable that the films used in protected cultivation have an increased life, thus producing less waste per annum. However, sustainability requires that a degradable material breaks down completely by natural processes so that the basic building blocks can be used again by nature to make a new life form. Plastics made from petrochemicals are not a product of nature and cannot be broken down by natural processes. It is assumed that the breakdown products will eventually biodegrade. In the meanwhile, these degraded, hydrophobic, high surface area plastic residues migrate into the water table and other compartments of the ecosystem causing irreparable harm to the environment [25].

Mechanisms of Degradation

Degradation is a complicated non-linear time-dependent process which affects directly or indirectly several properties of the material related to its functional characteristics. In its final stage of degradation, a material does not meet its functional requirements and is easily prone to mechanical failure. As a practical rule, the useful life of a material is considered to be reached when its initial mechanical strength is reduced by 50%. There are several factors to monitor and criteria to define the degree of degradation. Not all properties are affected by degradation in the same way though. Thus, the elongation at break (expressed as a percentage) appears to be a more sensitive 'index' of degradation than the tensile strength, the stress at yield or the modulus of elasticity. In fact, the material becomes more brittle with degradation, so it cannot retain its initial elongation at break [7].

Degradation of polymers is induced by different external factors and mechanisms. Briefly, the various degradation types for polymers are the following:

- Thermal degradation occurs due to use or processing of polymers at high temperatures.
- Photo-induced degradation occurs when, on exposure to the energetic part of the sunlight, i.e. the ultraviolet (UV) radiation, or other high-energy radiation, the polymer or impurities within the polymer absorb the radiation and induce chemical reactions.
- Mechanical degradation occurs due to the influence of mechanical stress–strain.
- Ultrasonic degradation is the application of sound at certain frequencies which may induce vibration and eventually breaking of the chains.
- Hydrolytic degradation occurs in polymers containing functional groups which are sensitive to the effects of water.

- Chemical degradation occurs when corrosive chemicals, such as ozone or the sulphur in agrochemicals, attack the polymer chain causing bond breaking or oxidation.
- Biological degradation is specific to polymer with functional groups that can be attacked by microorganisms.

Environmental Management Issues

Landfill Option

Landfill not only takes up large areas of land but can also generate bio-aerosols, odours and visual disturbance and may lead to the release of hazardous chemicals through the escape of leachate from landfill sites. Organic breakdown following landfill disposal of biodegradable waste, including bioplastics, causes the release of greenhouse gases. Landfill of waste usually implies an irrecoverable loss of resources and land (since landfill sites can normally not be used post-closure for engineering and/or health risk reasons), and in the medium to long term, it is not considered a sustainable waste management solution [26].

Incineration Option

The environmental impacts of incinerating plastic waste (as for most solid wastes or fuels) can include some airborne particulates and greenhouse gas emissions. Plants that are compliant with the Waste Incineration Directive are not thought to have any significant environmental impact. However, in some circumstances, energy recovery of plastic waste in MSW incinerators can result in a net increase in CO_2 emissions due to substituted electricity and heat production [27]. Therefore, all incineration activities should be associated with suitable filter system trap for released toxic substances, where the incinerators operate in a way not to pollute the atmosphere, soil and groundwater. There will also be an environmental burden due to the disposal of ashes and slag. For example, flue gas cleaning residues often have to be disposed of as hazardous waste due to the toxicity of the compounds they absorb. The net societal cost or benefit would of course depend on the alternatives, e.g. the existing power generation mix and the risk of open-air burning or landfill fires.

Recycling Option

In western countries, plastic consumption has grown at a tremendous rate over the past two or three decades. In the consumer societies of Europe and America, scarce petroleum resources are used for producing an enormous variety of plastics for an even wider variety of products. Many of the applications are for products with a life cycle of less than 1 year and then the vast majority of these plastics are then discarded. In most instances, reclamation of this plastic waste is simply not economically viable. In the industry (the automotive industry for example), there is a growing move towards reuse and reprocessing of plastics for economic as well as environmental reasons with many praiseworthy examples of companies developing technologies and strategies for recycling of plastics. Plastic recycling needs to be carried out in a sustainable manner. However, it is attractive due to the potential environmental and economic benefits it can provide. There is a wide variety of recycled plastic applications, and the market is growing.

However, the demand depends on the price of virgin material as well as the quality of the recycled resin itself. The use of recycled plastics is marginal compared to virgin plastics across all plastic types due to a range of technological and market factors. Recycled plastics are not commonly used in food packaging (one of the biggest single markets for plastics) because of concerns about food safety and hygiene standards, though this is beginning to change. Another constraint on the use of recycled plastics is that plastic processors require large quantities of recycled plastics, manufactured to strictly controlled specifications at a competitive price in comparison to virgin plastic. Such constraints are challenging, in particular, because of the diversity sources and types of plastic waste and the high potential for contamination. Not only is plastic made from a non-renewable resource but it is also generally non-biodegradable (or the biodegradation process is very slow). This means that plastic litter is often the most objectionable kind of litter and will be visible for weeks or months, and waste will sit in landfill sites for years without degrading.

Although there is also a rapid growth in plastic consumption in the developing world, plastic consumption per capita in developing countries is much lower than in industrialised countries. These plastics are, however, often produced from expensive imported raw materials. Not all plastics are recyclable. There are four types of plastics which are commonly recycled:

- Polyethylene - both high density and low-density polyethylene
- Polypropylene
- Polystyrene
- Polyvinyl chloride

A common problem with recycling plastics is that plastics are made up from parts of more than one kind of polymer or there may be some sort of fibre added to the plastic (a composite) to give added strength. This can make recovery difficult. When thinking about setting up a small-scale recycling enterprise, it is advisable to first carry out a survey to ascertain the types of plastics available for collection, the type of plastics used by manufacturers (who will be willing to buy the reclaimed material) and the economic viability of collection. The method of collection can vary. The following gives some ideas:

- House-to-house collection of plastics and other materials (e.g. paper)
- House-to-house collection of plastics only (but all types of polymer)
- House-to-house collection of certain objects only
- Collection at a central point, e.g. market or church
- Collection from street boys in return for payment
- Regular collection from shops, hotels, factories etc.

- Purchase from scavengers on the municipal dump
- Scavenging or collecting oneself

The method will depend upon the scale of the operation, the capital available for the set-up, transport availability etc. It should be noted that the ideas above assume an expansion in recycling capacity, which will require associated expansion in collection activities, use of secondary plastic materials and, associated with the latter, better methods for separating the different types of plastic to reduce contamination levels. These will allow the delivery of higher quality plastic waste streams to facilitate higher levels of recycling and to ensure quality markets for the secondary raw materials that result. Figure 7 is an example of the life cycle of recycled waste.

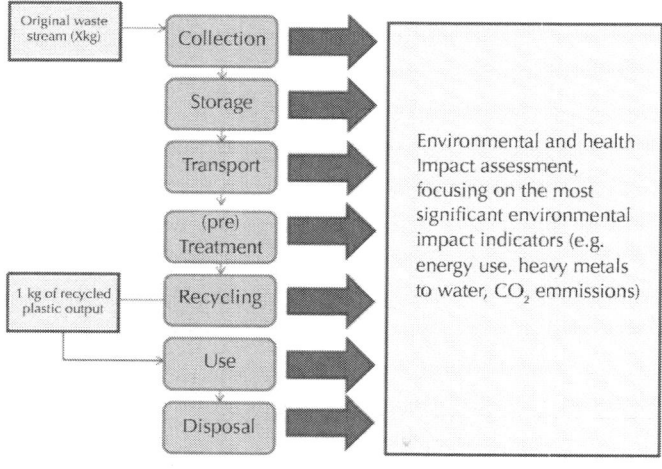

Figure 7: Life cycle approach for analysing the environmental impacts of plastic recycling.

Recycled PET

Post-consumer PET is often an attractive material for recycling. Unlike other polymers, nowadays, recycled PET can be produced and then directly suitable for contact with food if it is submitted to further decontamination steps such as super clean processes, which are able to decontaminate post-consumer contaminants to concentration levels of virgin PET materials [28]. PET can also be used in applications such as carpet fibres, geo-textiles, packaging and fibre fill. PET can be converted into polybutylene terephthalate (PBT) resin, which can be a valuable material for injection and blow-moulding applications. PBT is created through chemical polymerisation which converts the PET molecular chain into 'small repeating units', and through additional catalyst-assisted processes, PBT is produced. The polymerised PBT contains approximately 60% of the original mass of PET and can reduce solid waste by up to 900 kg for each tonne of PBT produced. Making PBT from recycled PET is often less energy consuming than producing the resin directly from oil stock (at 50 to 20 GJ/tonne, respectively) [29]. The main trends of interest in terms of economic impacts are anticipated to be the relative expansion of the recycling sector and questions regarding the economic impact of potentially lower economic growth on plastic waste treatment and secondary raw material use. The main social impacts are anticipated to be associated with health, and in particular, the epidemiological impacts are associated with the treatment of waste in third countries and the social perceptions around the continued use and increasing levels of plastic consumption and waste production.

Re-use Option

Products could be designed for re-use by facilitating the dismantling of products and replacement of parts. This could involve standardising parts across manufacturers. For example, LED lamp designs could benefit from standardisation of parts (many of which are plastic) to facilitate disassembly and remanufacturing [30].

Environmental Impacts

Pollutants released from burning plastic waste in a burn barrel are transported through the air either short or long distances and are then deposited onto land or into bodies of water. A few of these pollutants such as mercury, PCBs, dioxins and furans persist for long periods of time in the environment and have a tendency to bio-accumulate, which means they build up in predators at the top of the food web. Bioaccumulation of pollutants usually occurs indirectly through contaminated water and food rather than breathing the contaminated air directly. In wildlife, the range of effects associated with these pollutants includes cancer, deformed offspring, reproductive failure, immune diseases and subtle neurobehavioral effects. Humans can be exposed indirectly just like wildlife, especially through consumption of contaminated fish, meat and dairy products.

Environmental pollution can also be harmful to the structural integrity of the polyethylene due to chemical attack of the polymer bonds. Atmospheric pollutants such as nitrogen oxides, sulphur oxides, hydrocarbons and particulate can enhance the degradation of the polymers especially when combined with applied stress and must also be taken into account [31]. For instance, infrared studies have revealed that polyethylene reacts with NO_2 at elevated temperatures and that chemical attack is observed even at 25°C, probably due to the presence of olefinic bond impurities which react readily with NO_2. Similarly, SO_2 is rather reactive, especially in the presence of UV irradiation, which it readily absorbs and forms triplet excited sulphur dioxide ($^3SO_2*$). This species is capable of abstracting hydrogen from the polymer chains leading to the formation of macroradicals in the polymer structure, which in turn can undergo further depolymerisation.

Overall, the level of environmental impact associated with plastic waste is anticipated to increase over the period until 2015 due to continued growth in plastic waste production (associated with continued rises in plastic waste consumption). More specifically, greenhouse gas emissions associated with the plastic life cycle are anticipated to increase, albeit on a lower trajectory than in the past. Negative consequences in terms of littering and plastic pollution in marine waters would also be anticipated to increase in the absence of any additional curbs [4]. Due to many factors, not the least of which is their ready availability, 96% of all plastic grocery bags as an example are thrown into landfills [32]. However, plastic bags decompose very slowly, if at all. In fact, a bag can last many years, inhibiting the breakdown of biodegradable materials around or in it [32].

Lightweight plastic grocery bags are additionally harmful due to their propensity to be carried away on a breeze and become attached to tree branches, fill roadside ditches or end up in public waterways, rivers or oceans. In one instance, Cape Town, South Africa had more than 3,000 plastic grocery bags that covered each kilometre of the road [32]. In this century, an estimated 46,000 pieces of plastic are floating in every square kilometre of the ocean worldwide [32].

What is most distressing is that over a billion seabirds and mammals die annually from ingestion of plastics [32]. According to UNEP, plastic waste in the ocean causes the deaths of up to one million seabirds, 100,000 marine mammals and countless fish every year [33]. Big animals (e.g. turtles, whales, seals, and sea lions) can be trapped by nets and films and eat the small particles of plastics which may block their digestive systems. Entanglement and ingestion of plastic fragments can even lead to death by drowning, suffocation, strangulation or starvation through reduced feeding efficiency. At least 267 different species are known to have suffered from entanglement or ingestion of marine debris, including seabirds, turtles, seals, sea lions, whales and fish [34]. According to Brown, in Newfoundland, 100,000 marine mammals are killed each year by ingesting plastic [35]. However, the impact of plastic bags does not end with the death of one animal; when a bird or mammal dies in such a manner and subsequently decomposes, the plastic bag will again be released into the environment to be ingested by another animal.

Another environmental issue involves blowing agents used to make foamed plastics. All blowing agents eventually escape to the atmosphere, and among them, there is a particular concern with the CFC stratospheric ozone layer. An international treaty was signed in 1990 on CFCs which was completely implemented in 2000. In some more restricted geographical area, the smoke-forming potential of hydrocarbon blowing agents is also considered an issue.

Because CFCs have been widely employed in foamed plastics, including polystyrene, rigid and flexible polyurethanes and polyolefins, for example, there has been intense activity to develop satisfactory substitutes. Among the most promising of these are the HCFCs, which have only 2% to 10% of the ozone depletion potential of CFCs, and hydrofluorocarbons, which have zero ozone depletion potential. The current consensus is that the HCFCs represent a transitional solution to the problem. There has been promising development work with CO_2 as a blowing agent for polystyrene foam sheet. In this application, CO_2 is considered environmentally benign.

Dioxins

Dioxins are unintentionally but unavoidably produced during the manufacture of materials containing chlorine, including PVC and other chlorinated plastic feedstocks. Burning these plastics can release dioxins. Polychlorinated dibenzo-p-dioxins and polychlorinated dibenzofurans, hexachlorobenzene and PCBs are unintentional persistent organic pollutants (U-POPs), formed and released from thermal processes involving organic matter and chlorine as a result of incomplete combustion or chemical reactions. These U-POPs are commonly known as dioxins because of their similar structure and health effects [36].

Dioxin is a known human carcinogen and the most synthetic carcinogen ever tested in laboratory animals. A characterization by the National Institute of Standards and Technology of cancer-causing potential evaluated dioxin as over 10,000 times more potent than the next highest chemical (diethanol amine), half a million times more than arsenic and a million or more times greater than all others. Also, open burning of plastic or incineration involves air emissions of sulphur dioxide, nitrogen dioxide, chlorine, dioxin and fine particulates and emissions of greenhouse gases of CO_2 and nitrous oxide (N_2O); ash which remains after incineration needs to be disposed of and can be toxic.

The pyrolysis or combustion of even a simple synthetic polymer produces mix grill of gases. Most of these gases are self-toxic, i.e. interfering with the normal biochemical processes of the body or exclude O_2 from the victim. It must be understood that the type and the concentration of these gases in any fire situation vary from material to material as well as the environment. A few of them together with their physiological effects are shown in Table 4.

Table 4: Physiological effects of some gases involved in combustion

Gas concentration	Effect in all fire conditions
Oxygen (O_2 (%))	
21	Normal concentration in air
2 to 15	Shortness of breath, headache, dizziness, quickened pulse, fatigue on exertion, loss of muscular coordination for skilled movement
10 to 12	Nausea and vomiting, exertion impossible, paralysis of motion
6 to 8	Collapse and unconsciousness, but rapid treatment can prevent death
6	Death in 6 to 8 min
2 to 3	Death in 45 s
Carbon monoxide (CO (ppm))	
400	Nausea after 1 or 2 h, collapse after 2 h and death after 3 to 4 h
1,000	Difficulty in ambulation, death after 2 h
2,000	Death after 45 min
3,000	Death after 30 min
5,000	Rapid collapse, unconsciousness and death within a few minutes
Carbon dioxide (CO_2 (ppm))	
250 to 350	Normal concentration in air
25,000	Ventilation increased by 100%
50,000	Symptoms of poisoning after 30 min, headache, dizziness and sweating
120,000	Immediate unconsciousness, death in a few minutes

Hydrogen cyanide (HCN (ppm))	
45 to 54	Tolerable for 1/2 to 1 h[a]
110 to 135	Fatal after 1/2 to 1 h[a]
181	Fatal after 10 min[a]
280	Immediately fatal[a]
Hydrogen chloride (HCl (ppm))	
5 to 10	Mild irritation of mucus membranes[b]
50 to 100	Barely tolerable[b]
1,000	Danger of lung oedema after a short exposure[b]
1,000 to 2,000	Immediately hazardous to life[b]
Acrolein (CH_2CHCHO (ppm)) (PVC)	
1	Immediately detectable irritation[c]
5.5	Intense irritation[c]
<10	Lethal in short time[c]
24	Unbearable[c]

[a]Effect from nitrogen-containing polymeric materials, e.g. acrylics, wool, urethane foam etc.; [b]effect from chloride-containing polymers e.g. PVC; [c]effect from many polymeric materials, e.g. wool and polypropylene.

Nkwachukwu et al.

Nkwachukwu et al. International Journal of Industrial Chemistry 2013 4:34 doi: 10.1186/2228-5547-4-34.

Health Effects

Because of the persistent and bio-accumulative nature of dioxins and furans, these chemicals exist throughout the environment. Human exposure is mainly through consumption of fatty foods, such as milk. IPEP [36] notes that 90% to 95% of human exposure to dioxins is from food, particularly meat and dairy products. Their health effects depend on a variety of factors, including the level of exposure, duration of exposure and stage of life during exposure. Some of the probable health effects of dioxins and furans include the development of cancer, immune system suppression, reproductive and developmental complications and endocrine disruption [36]. The International Agency for Research on Cancer has identified 2, 3, 7, 8-tetrachlorodibenzodioxin as the most toxic of all dioxin compounds.

High exposure to PDBEs, which accumulate in the human body, has been linked to thyroid hormone disruption, permanent learning and memory impairment, behavioural changes, hearing deficits, delayed puberty onset, impaired infant neurodevelopment, and decreased sperm count, fetal malformations and possibly cancer. These activities lead to severe pollution of soils by POPs and heavy metals in the countries concerned, which may also affect the surrounding environment such as rice fields and rivers via atmospheric movement and deposition [37-39].

Toxic emissions produced during the extraction of materials for the production of plastic grocery bags, their manufacturing and their transportation contribute to acid rain, smog and numerous other harmful effects associated with the use of petroleum, coal and natural gas, such as health conditions of coal miners and environmental impacts associated with natural gas and petroleum retrieval [40]. Heavy metals may be released into the environment from metal smelting and refining industries, scrap metal, plastic and rubber industries, and various consumer products and from burning of waste containing these elements. On release to the air, the elements travel for large distances and are deposited onto the soil, vegetation and water depending on their density. Once deposited, these metals are not degraded and persist in the environment for many years poisoning humans through inhalation, ingestion and skin absorption. Acute exposure leads to nausea, anorexia, vomiting, gastrointestinal abnormalities and dermatitis.

Impacts on human health are perhaps the most serious of the effects associated with plastic grocery bags, ranging from health problems associated with emissions to death. In the year 2005, the city of Mumbai, India experienced massive monsoon flooding, resulting in at least 1,000 deaths, with additional people suffering injuries [32]. City officials blamed the destructive floods on plastic bags which clogged gutters and drains, preventing the rainwater from leaving the city through underground systems. Similar flooding happened in 1988 and 1998 in Bangladesh, which led to the banning of plastic bags in 2002 [40]. By clogging sewer pipes, plastic grocery bags also create stagnant water; stagnant water produces the ideal habitat for mosquitoes and other parasites which have the potential to spread a large number of diseases, such as encephalitis and dengue fever, but most notably malaria.

CONCLUSIONS

Overall, the level of environmental impact associated with plastic waste is anticipated to increase over the period until 2015 due to continued growth in plastic waste production (associated with continued rises in plastic waste consumption). Over this period, the rise in environmental impacts is anticipated to be comparatively slower than in the past as much of this increase in production is dealt with by recycling and energy recovery expansion. However, disposal levels are only anticipated to remain static or drop in a limited way, maintaining the overall picture of the environmental footprint.

In terms of environmental impacts, the following trends are considered to be of most significance:

Rising use of plastics. The primary plastic feedstock will remain fossil fuels, despite the anticipated rapid rise in the production of bioplastics. This implies continued reliance on carbon-intensive production methods, with relatively high levels of embodied carbon and energy in the products. While traditional refineries might be driven to be more efficient over the projection period due to changes in rules surrounding for example the Fuel Quality Directive (which requires life cycle reductions in transport fuels), such efficiencies are likely to be offset by the increasing the level of production and demand.

Rising levels of plastic waste generation. This implies the need for an expanded waste management system simply to remain capable of dealing with the anticipated increase waste production.

Increasing levels of recycling. Recycling rates are anticipated to increase over the outlook period, and end markets are developing. However, the proportion of disposal is expected to remain significant. This implies a significant expansion in the overall Mt amount of waste recycled, i.e. a similar proportion of a greater quantity of waste will be recycled. This in turn implies three key evolutions in the plastic waste recycling business. Firstly, an expansion in the collection of plastic waste, secondly an expansion in processing capacity and thirdly an expansion in the use of secondary plastic materials. Legislation has already been proposed or passed in the USA of the federal, state and local levels restricting or banning the use of some plastics in applications where there is perceived problem. This trend is sure to continue. The technical community has not lagged in developing responses to these challenges expect in the developing nations. The principal routes being followed are degradation, incineration and recycling. Another important approach which involves both consumers and manufacturers is source reduction. The activity in this area is focused largely on warning consumers from the throwaway habits that have developed over recent decades, particularly with packaging waste stream. These efforts are evolving so rapidly that it is difficult to predict how each one will impact the problem over the longer term. European approaches. In Europe, the principal measures implemented to deal with plastics are the producer responsibility mechanisms - these do not target plastic bags specifically but aim to encourage the recycling and recovery of plastics. Different member states use different approaches, but in most countries, the packaging industry makes payments to designated bodies that are responsible for arranging the collection, separation, recycling and recovery of a predetermined amount of packaging. A notable feature is that these fees paid by the packaging industry are not necessarily passed on to consumers in a transparent manner. Therefore, the collection of taxes and public awareness can reduce indiscriminate use of plastic bags.

Recommendation

The redesign of plastic products, both at the scale of the individual polymer and in terms of the product's structure, could help alleviate some of the problems associated with plastic waste. With thoughtful development, redesign could have an impact at all levels of the hierarchy established by the European Waste Framework Directive: prevention, re-use, recycle, recovery and disposal [1]. Infrastructure for the safe disposal and recycling of hazardous materials and municipal solid waste should be developed. Approximately 50% of all waste is organic and can therefore be composted. Another large segment of the remainder can be recycled, leaving only a small portion to be disposed. The remaining portion can then be disposed through sanitary landfills, sewage treatment plants and other technologies.

In general, disposal via modern, sanitary landfills is not very effective with biodegradable plastic materials because the limited availability of oxygen and moisture retards most biodegradation processes. As yet, bioplastics cannot replace all types of plastic, particularly certain types of food packaging that require gas permeability. Biodegradable plastics are most effectively treated in compositing systems where aerobic processes predominate. Biodegradation may also influence the types and concentrations of soil microflora in disposal areas. Enrichment of soil with certain microflora could have unanticipated risks, such as an outbreak of a new microbial disease [41].

At present, there is a growing consensus that the concept of degradable plastics has been oversold as a solution to the waste disposal problem, primarily because a large portion of degradable plastics will end up in landfills where breakdown tends to be very slow. The most promising applications of degradable plastics probably are for more limited problems, such as litter, where sunlight, air, moisture and microorganism are generally available to accelerate polymer breakdown [42].

High-temperature incineration of waste plastics has at least two potential advantages. First, increasingly scarce landfill space is conserved because the volume reduction from feed to ash with a well-operated incinerator is in 90% range or higher; second, the high generation of stream, electricity or both. Incinerators have drawbacks, however. They can emit corrosives such as HCl and traces of highly toxic dioxins and furans if chlorine-containing materials such as PVC or bleached paper are in the feed stream. These emissions could pose hazards, especially to people living close to the incinerator; compounds of toxic heavy metals such as lead, chromium and cadmium are present in some plastic products as stabilizers and pigments, and these elements tend to end up concentrated in the ash. If the compounds of these metals in the ash are leachable, soil and groundwater contamination is possible. Advocates of incineration are convinced; however, that current technology will prevent stack discharge of most toxic and corrosive combustion products. They further claim that only a very small fraction of the heavy metals in the ash is leachable, and this should pose no problem if proper disposal procedures are followed. Moreover, heavy metal-based pigments and stabilizers gradually are being phased out of many applications in favour of organic substitutes. Current trends suggest increasing reliance on the incineration approach despite the claimed drawbacks.

Recycling has obvious environmental advantages and is not opposed by any strong voting blocs or economic interests. Recycling of plastics should be carried in a manner to minimize pollution during the process and enhance efficiency and conserve the energy. Much of the future success of plastic waste recycling will depend on the development of effective collection and separation systems, which, along with appropriate incentives, will ensure the broad and willing participation of industry and consumers. In this case, involving pre-consumer waste or scrap, where the material identifies and uniformity can be reasonably maintained, it is often possible to recycle the plastic back to the same product. In other instances, including those where carefully targeted post-consumer collection methods are possible. A secondary product can be made of a single recycled plastic, for example, PET beverage bottle scrap can be recycled to fabricate bottles for non-beverage applications [7]. Clearer certification and labelling schemes are needed to ensure that the public understands what is meant by biodegradable, compostable or eco-friendly. DG

Environment's report on Plastic waste in the environment [11] proposed that any targets on bioplastics should be combined with a labelling system and initiatives to increase public awareness and education. Labelling of plastic parts with the type of polymer they contain could also help in sorting for recycling and re-use. Along with the plastic waste issue, significant new developments can be anticipated as the industry continues its aggressive search for solutions to these important environmental problems. The redesign of plastics and bioplastics has the potential to reduce the use of fossil fuels, decrease CO_2 emissions and decrease plastic waste. There is a need to increase public awareness through litter education as an important supporting element and other initiatives that may be undertaken to reduce plastic waste and their impacts.

AUTHOR'S CONTRIBUTIONS

This work was finished through the collaboration of all authors. OIN conceived the study and drafted the manuscript together with AOI. CHC and LA carried out the computations in the manuscript. CHC, AOI and LA participated in the coordination and in revising the manuscript. All authors read and approved the final manuscript.

ACKNOWLEDGMENTS

The authors would like to thank all the reviewers who read this paper carefully and provided valuable suggestions and comments. The editorial assistance of our colleagues Mr. Kanno Okechukwu Charles and Mr. Chukwuma Royal are very much appreciated.

REFERENCES

1.   Science for Environmental Policy (2011) Plastic waste: redesign and biodegradability Brussels: European Commission pp 1–8

2. Thompson RC, Swan SH, Moore CJ, Vom Saal FS (2009) Our plastic age. Philosophical Transactions of the Royal Society. 364:1973–1976.

3. Plastics Europe (2007) The compelling facts about plastics - an analysis of plastics production, demand and recovery for 2005 in Europe. Brussels: Plastics Europe. http://www. plasticseurope.org/Content/Default.asp?PageID=517#

4. Mudgal S, Lyons L, Bain J, Débora D, Thibault F, Linda J (2011) Plastic waste in the environment: revised final report. European Commission DG Environment. France: Bio Intelligence Service. http://www. ec.europa.eu/environment/ waste/studies/pdf/plastics.pdf webcite. Accessed April 2011

5. Europe P (2009) The compelling facts about plastics - an analysis of European plastics production, demand and recovery for 2008 Brussels: Plastics Europe

6. Hopewell J, Dvorak R, Kosior E (2009) Plastics recycling: challenges and opportunities. Philosophical Transactions of the Royal Society. 364:2115–2126

7. Onwughara IN, Nnorom IC, Kanno OC, Chukwuma RC (2010) Disposal methods and heavy metals released from certain electrical and electronic equipment wastes in Nigeria: adoption of environmental sound recycling system. International Journal of Environmental Science and Development 1(4):290–296

8. Balakrishnan RB, Anand KP, Chiya AB (2007) Electrical and electronic waste: a global environmental problem. Journal of Waste Management and Research. 25:307–317

9. Antrekowitsch H, Potesser M, Spruzina W, Prior F (2006) Metallurgical recycling of electronic scrap San Antonio pp 899–904

10. Plasticsnews   (http://plasticsnews.com/fyi-charts/index. html?id= 17731. Accessed 13 October 2008) Paying more for less. http://plasticsnews. com/fyi-charts/index.html?id=17731 webcite. Accessed 13 October 2008

11.   Mudgal S, Lyons L (2010) Plastic waste in the environment: final report France: Bio Intelligence Service

12.   Plastics Europe (2010) Plastics - the facts. An analysis of European plastics production, demand and recovery for 2009. http://www. plasticseurope.org/document/plastics---the-facts-2010.aspx?FolID=2 webcite. Accessed 27 October 2010

13.   Association of Plastics Manufacturers in Europe (APME) (2000) An analysis of plastics consumption and recovery in Europe Brussels: APME

14.   Thaniya K (2009) Sustainable solutions for municipal solids waste management in Thailand. World Academy of Science, Engineering and Technology 36:666–671

15.   Onwughara IN, Nnorom IC, Kanno OC (2010) Issues of roadside disposal habit of municipal solid waste, environmental impacts and implementation of sound management practices in developing country: "Nigeria". International Journal of Environmental Science and Development 1(5):409–417

16.   Luitel KP, Khanal SN (2010) Study of scrap waste in Kathmandu Valley. Kathmandu University Journal of Science, Engineering and Technology 6(1):116–122

17.   Chi JO, Sung OL, Hyung SY, Tae JH, Myong JK (2003) Selective leaching of valuable metals from waste printed circuit boards. J Air Waste Manage Assoc. 53:897–898

18.   Witten E (2009) The composites market in Europe Germany: AVK

19.   Marcos NC (2010) Renewable chemicals and polymers. What's next? SRI Consulting 2010. http://www. apla.com.ar/img/conferencias/84_conf.pdf webcite. Accessed November 2010

20.   European Bioplastics (2011) European Bioplastics, driving the evolution of plastics. http://european-bioplastics. org/index. php?id=141

21. Kurudufu P (2009) Recycling plastic. Practical action Eastern Africa. http://practicalaction. org/docs/technical_ information_service/recycling_plastics.pdf webcite. Accessed 2 October 2009

22. Zereena BI (2010) Plastics and environment. Dissemination Paper - 12. Centre of Excellence in Environmental Economics. http://coe. mse.ac.in/dp/Paper12.pdf webcite. Accessed 20 April 2010

23. Tippayawong NN, Khongkrapan P (2009) Development of a laboratory scale air plasma torch and its application to electronic waste treatment. International Journal for Environmental Science and Technology 6(3):407–411

24. Lardinois I, Van de K (1995) A plastic waste, option for small-scale resource recovery Amsterdam: TOOL

25. Gautam SP (2009) Bio-degradable plastics - impact on environment. Central Pollution Control Board Ministry of Environment and Forests Government of New Delhi: India

26. Commission E (2008) Green Paper on the management of bio-waste in the European Union European Commission, Brussels: COM

27. Pilz H, Brandt B, Fehringer R (2010) The impact of plastics on life cycle energy consumption and greenhouse gas emissions in Europe Plastics Europe, Brussels: Denkstatt summary report

28. Welle F (2011) Twenty years of PET bottle to bottle recycling— an overview. Resources, Conservation and Recycling. 55(11):865–875

29. Plastemart (2003) Green method of manufacturing virgin PET/PBT from recycled products offers energy saving. http://plastemart.   com/upload/Literature/Green-method-manufacture-virgin%20PET-PBT-recycled-products-energy%20saving-Valox%20iQ-Xenoy%20iQ.asp   webcite. Accessed 15 May 2009

30. Hendrickson CT, Matthews DH, Ashe M, Jaramillo P, McMichael FC (2010) Reducing environmental burdens of solid-state lighting through end-of-life design. Environ Res Lett.

31. Dilara PA, Briassoulis D (2000) Degradation and stabilization of low-density polyethylene films used as greenhouse covering materials. Journal of Agricultural Engineering Resource. 76:309–321 http://www. idealibrary.com

32. Ellis S, Kantner S, Saab A, Watson M (2005) Plastic grocery bags: the ecological footprint Victoria: VIPIRG pp 1–19.

33. UNEP (2006) Ecosystems and biodiversity in deep waters and high seas. Switzerland. http://unep. org/pdf/EcosystemBiodiversity_DeepWaters_20060616.pdf

34. Derraik JGB (2002) The pollution of the marine environment by plastic debris: a review. Mar Pollut Bull. 44:842–852.

35. Brown S (2003) Seven billion bags a year. Habitat Australia 31(5):P28

36. The International POPs Elimination Project (IPEP) (2005) A study on waste incineration activities in Nairobi that release dioxin and furan into the environment Kenya

37. Wong MH, Wu SC, Deng WJ, Yu XZ, Luo Q, Leung AOW, Wong CSC, Luksemburg WJ, Wong AS (2007) Export of toxic chemicals - a review of the case of uncontrolled electronic-waste recycling. Environ Pollut. 149:131–140.

38. Environmental Working Group (2003) Mother's milk - toxic fire retardants (PBDEs) in human breast milk. Washington, DC: Environmental Working Group.

39. Herbstman JB, Andreas S, Matthew K, Sally AL, Richard SJ, Virginia R, Larry LN, Deliang T, Megan N, Richard YW, Frederica P (2010) Prenatal exposure to PBDEs and neurodevelopment. Environ Health Perspect. 118(5):712–719.

40. Environmental Literacy Council (2005) Paper or plastic?. http://www. enviroliteracy.org/article.php/1268.html webcite. Accessed 20 November 2005

41. Sudesh K, Iwata T (2008) Sustainability of biobased and biodegradable plastics. Clean 36(5–6):433–442
42. Taylor DA (2010) Principles into practice: setting the bar for green chemistry. Environ Health Perspect. 118(6):254–257

Engineering Aspects and Practical Application of Autotrophic Nitrogen Removal from Nitrogen Rich Streams

Stijn W.H. Van Hulle[a, b], Helge J.P. Vandeweyer[b],
Boudewijn D. Meesschaert[c],
Peter A. Vanrolleghem[a, d], Pascal Dejans[b], and Ann
Dumoulin[b]

[a]BIOMATH, Department of Applied Mathematics, Biometrics and
Process Control, Ghent University, Coupure Links 653, B-9000
Gent, Belgium

[b]Research Group EnBiChem, Department of Industrial Engineering
and Technology, University College West Flanders, Graaf Karel de
Goedelaan 5, B-8500 Kortrijk, Belgium

cChemistry Department, Catholic University College of Bruges–Ostend, Zeedijk 101, B-8400 Ostend, Belgium

dModelEAU, Département de génie civil, Pavillon Pouliot, Université Laval, Québec G1K 7P4, QC, Canada

ABSTRACT

The anaerobic ammonium oxidation (Anammox) process, discovered 20 years ago, is, in combination with partial nitritation, ideally suited to treat nitrogen rich waste water streams such as digester effluent. In this review the engineering aspects and the practical application of the process are reviewed. The conventional nitrification–denitrification and nitritation–denitritation are also discussed briefly.

The environmental conditions affecting the nitrification process, free ammonia and nitrous acid concentration, temperature, pH and dissolved oxygen concentration, are discussed. These conditions can be controlled in such a way that the partial nitritation step produces an Anammox-suited influent. The Anammox reactor conditions should favour the growth of the Anammox organisms in view of their low growth rate and possible inhibition effects. Dissolved oxygen and nitrite concentrations should be kept as low as possible and biomass washout should be limited. If the partial nitritation process and the Anammox process are occuring in the same reactor, care should be taken to the dissolved oxygen concentration, the ammonium load and the nitrite concentration to obtain a sustainable co-existence between aerobic and anaerobic ammonium oxidizers.

An overview is presented of the practical implementation of autotrophic nitrogen removal. The process can be accomplished in the same reactor (1-reactor system) or by using 2 separate reactors (2-reactor system). Typically the 1-reactor system is a biofilm or granular reactor where the ammonium oxidizers are active in the outer layers of the biofilm or granule, producing a suitable amount of nitrite for the Anammox organisms that are active in the inner layers. Transport of ammonium and the produced nitrite is governed by diffusion. Finally, the different nitrogen removal processes are compared in terms of operational conditions and a direction for future work is provided.

INTRODUCTION

One of the elements of concern in wastewater is nitrogen, especially since the use of synthetic nitrogen fertilizer produced from atmospheric N_2 by the Haber–Bosch process has increased tenfold over the last 40 years. The human contribution to nitrogen pollution, for example in the form of urine, is ever increasing in view of the growing world population. Discharge of this nitrogen into the natural waters can lead to, amongst others, eutrophication and oxygen depletion.

In most modern wastewater treatment plants (WWTP) nitrogen, which is generally in the form of ammonium or organic nitrogen, is removed by biological nitrification/denitrification (reaction (3)). As a first step ammonium is converted to nitrate (nitrification, reaction (1)) which is then, in a second step, converted to nitrogen gas (denitrification, reaction (2)). Benefits of the process are the high potential removal efficiency, high process stability and reliability, relatively easy process control, low area requirement and moderate cost [1]

$$NH_4^++2O_2+2HCO_3^-\rightarrow NO_3^-+3H_2O+2CO_2 \tag{1}$$
$$5C+4NO_3^-+2H_2O\rightarrow CO_2+4HCO_3^-+2N_2 \tag{2}$$
$$4NH_4^++8O_2+5C+4HCO_3\rightarrow 2N_2+10H_2O+9CO_2 \tag{3}$$

Generally, the conventional biological nitrogen removal process is used for treating wastewaters with relatively low nitrogen concentrations (total nitrogen concentration less than 100 mg N L^{-1}). Some wastewater streams such as anaerobic digester effluents, landfill leachate and industrial wastewaters contain high concentrations of nitrogen [2]. One more sustainable alternative is the nitritation–denitritation process over nitrite (Eqs. (4) and

(5)). This process requires less oxygen and less organic carbon in comparison with the traditional nitrification–denitrification.

$$NH_4^+ + 1.5O_2 + 2HCO_3^- \rightarrow NO_2^- + 3H_2O + 2CO_2 \tag{4}$$

$$3C + 4NO_2^- + 2H_2O + CO_2 \rightarrow 4HCO_3^- + 2N_2 \tag{5}$$

The ANaerobic AMMonium OXidation (Anammox) process, which was discovered about 20 years ago [3] but was already predicted to exist 30 years ago [4], could offer another alternative for the treatment of this return stream.

In the Anammox process ammonium is oxidized under anoxic, i.e. oxygen depleted, conditions with nitrite as electron acceptor (Eq. (6)). Ammonium and nitrite are consumed on an almost equimolar basis. The Anammox process should always be combined with a partial nitritation process, such as the SHARON process [5], where half of the ammonium is oxidized to nitrite. Both autotrophic processes will increase the sustainability of wastewater treatment as the need for carbon addition (and concomitant increased sludge production) is omitted and oxygen consumption and the emission of nitrous oxide during oxidation of ammonia are largely reduced [6]. Especially since nitrous oxide has become a significant factor in the greenhouse gas footprint of the total water chain [7]

$$NH_4^+ + 01.32NO_2^- + 0.66HCO_3^- \rightarrow 1.02N_2 + 2.03H_2O + 0.66CH_2O_{1.5}N_{0.15} + 0.26NO_3^- \tag{6}$$

The combined process (partial nitritation and Anammox) was termed autotrophic nitrogen removal process and is depicted in Fig. 1.

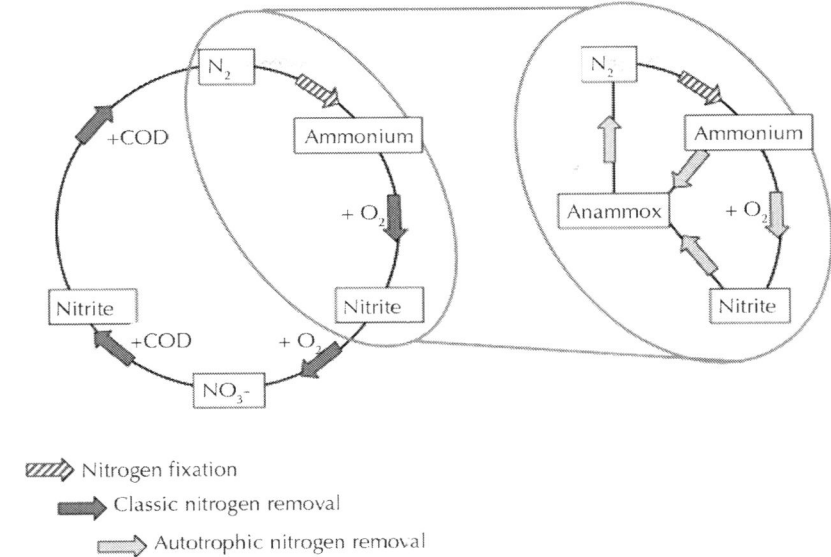

Figure 1: The updated nitrogen cycle with autotrophic nitrogen removal.

In this contribution the engineering aspects and practical application of this autotrophic nitrogen removal process will be reviewed, with a main focus on the Anammox process.

NITROGEN ELIMINATION FROM WASTEWATER BY AUTOTROPHIC NITROGEN REMOVAL

For a specific application the available alternatives for nitrogen elimination need to be evaluated on a multitude of cost aspects, chemical and energy requirements, operational experience, process reliability and environmental impact. However, the selection of the best alternative is generally based on cost-effectiveness. Nitritation–denitritation produce savings in oxygen demand during nitrification, a reduction of organic matter requirements in the denitrification process and a decrease in surplus sludge production.

The application of partial nitritation–Anammox goes even further in these 3 requirements as demonstrated in Table 1.

Table 1: Comparison of various nitrogen removal processes (adjusted from [8] and [9])

Process	Oxygen needed $(gO_2 \ g \ N^{-1})$	COD needed without assimilation $(gCOD \ g \ N^{-1})$	COD needed with assimilation $(gCOD \ g \ N^{-1})$
Nitrification–denitrification	4.57	2.86	4.0
Nitritation–denitritation	3.43	1.72	2.4
Partial nitritation–Anammox	1.72	–	–
OLAND	1.94	–	–

Compared to the traditional nitrogen removal process, which involves nitrification–denitrification, respectively 25% and 60% less oxygen will be consumed by the nitritation–denitritation and partial nitritation–Anammox process. Moreover, a lower (40% less) or no organic carbon source is needed if denitritation or Anammox is used as second step [8]. Nitrate produced by the Anammox process and by the OLAND process requires also organic carbon. However, most real wastewater contain a small amount of biodegradable COD which can be used to denitrify the produced nitrate. The three processes require more or less the same buffering capacity from the treated water: 1 mol H^+ is produced per mol N converted. This implies that all processes do not require significant pH control costs in the case of a sufficient buffer capacity in the waste water stream (1 mol HCO_3^- per 1 mol NH_4^+) [10]. Mulder [11] stated that the sludge production decreased from 1.0 to 0.1 g dry weight g^{-1} N when nitrification–denitrification is compared with OLAND/partial nitritation–Anammox. As a consequence, sludge treatment and disposal costs significantly decrease. Vlaeminck [10] calculated that the application of OLAND for treating reject water could save about 85% of the operational costs in comparison with nitrification/denitrification. The cost of the autotrophic nitrogen removal process is 1 euro per kg N removed, while other conventional nitrogen removing techniques cost 2–4 euro per kg N removed. As such, 1–3 euro per kg N removed can be saved [5].

In practice the selection of either a biological or a physiochemical method for nitrogen elimination is also determined by the nitrogen concentration of the wastewater. According to Mulder [11] biological treatment by autotrophic nitrogen removal is to be preferred for concentrated wastewater streams with ammonium concentrations in the range of 100–5000 mg N L^{-1}. For this concentration range, the autotrophic nitrogen process offers a sustainable alternative to conventional techniques, such as conventional nitrification and denitrification since less energy and chemicals are needed.

Typical wastewaters with high ammonia concentrations are reject water, piggery manure, landfill leachate and some industrial waste waters (examples of these wastewaters are depicted in Table 2). Reject water originating from dewatered sludge is normally recycled to the influent of the WWTP. An analysis of the nitrogen balance of the WWTP Dokhaven in Rotterdam (The Netherlands) revealed that the reject water accounted for only a few percentages of the total flow, but 15% of the nitrogen load [5] and [12]. A separate treatment of this stream reduces the nitrogen load coming from the digesters and can significantly help to reduce the nitrogen concentration in the effluent of the main wastewater treatment plant. This way ever reducing discharge limits for nitrogen concentrations can be met. The landfill leachate is heavily polluted and has to be captured and treated. Often this leachate is recycled over the landfill, decreasing the COD concentration and increasing the nitrogen concentration because the landfill acts as an anaerobic bioreactor [13]. Similar to reject water this landfill leachate is characterized by a high ammonium and a low COD content [14]. Raw manure is usually separated into a thick and a thin fraction. The thick fraction can be used as soil enhancer while the thin (liquid) fraction is treated. The composition of this thin fraction can vary and depends on the separation method and the composition of the animal feed [15]. However, large concentrations of COD, nitrogen and phosphorus can be expected, which is not favourable for autotrophic nitrogen removal. Industrial processes can generate streams with high nitrogen content. Also, concentrated industrial streams with high COD content can lead to highly loaded nitrogen streams if they are first treated in an anaerobic digester. Examples can be found in the pharmaceutical industry [16], tanneries [17], slaughterhouse waste processing [18], potato processing industries, alcohol and starch production [19] and formaldehyde production [20] and [21]. A shortcoming of a lot of these streams is the presence of recalcitrant and/or toxic components in the streams, resulting in a high effluent COD concentration. In the latter case, despite the favourable $C N^{-1}$ ratio there is still a need for carbon addition to allow the necessary denitrification. Carucci et al. [22] reported a minimum $C N^{-1}$ ratio of 8 for tannery wastewater, which is much higher than the normally applied ratio of 4–6 [11]. This tannery wastewater also contains chromium, sulphide and chloride, all resulting in negative effects on the nitrification process [17] and [23]. Wastewater of formaldehyde production, characterized by a high organic COD content, partially inhibits both nitrification and denitrification [21] and will thus lead to a more difficult operation.

(PARTIAL) NITRIFICATION

Nitrification is the aerobic oxidation of ammonium to nitrate. It consists of two sequential steps carried out by two phylogenetically unrelated groups of aerobic chemolithoautotrophic bacteria. Some heterotrophic bacteria can also oxidize ammonium to nitrate, but this is only a very small contribution to the overall ammonia oxidation. First, ammonium is oxidized to nitrite by the aerobic ammonia-oxidizing bacteria. Approximately 2 mol of protons are produced for every mole of ammonium oxidized. Ammonium oxidation is therefore an acidifying reaction. In the second step nitrite is oxidized to nitrate by the nitrite oxidizing bacteria. No single known autotrophic bacterium is capable of complete oxidation of ammonium to nitrate in a single step [24]. The key reactions of nitrification are given by Eqs. (7) and (8):

$$NH_4^+ + \tfrac{3}{2}O_2 \rightarrow NO_2^- + H_2O + 2H^+ \tag{7}$$

$$NO_2^- + \tfrac{1}{2}O_2 \rightarrow NO_3^- \tag{8}$$

In view of coupling partial nitritation with Anammox, nitrite oxidizing activity should be suppressed and ammonium should only be oxidized for about 50% to nitrite. Different influencing factors can be used to engineer a system that accomplishes this requirement, as discussed below.

The most important environmental parameters to obtain partial nitritation are the free ammonia (FA, NH_3) and free nitrous acid (FNA, HNO_2) concentration, the temperature, pH and dissolved oxygen concentration. The difference in sensitivity of ammonium and nitrite oxidizers towards these influences determines whether there will be nitrite accumulation in a nitrifying system. Indeed, nitrite oxidizers are generally more sensitive to detrimental environmental conditions than ammonium oxidizers. Hydroxylamine and chlorate inhibit irreversible nitrite oxidizers but not ammonium oxidizers. These compounds are able to inactivate the nitrite oxidizers while ammonia and nitrous acid can lead to adaptation of nitrite oxidizers. Furthermore, it is necessary to consider economic feasibility when using temperature, pH and inhibitor as regulation parameter. By applying lower oxygen concentrations, aeration will be saved and nitrite oxidizers cannot grow into the system [25].

An Anammox-suited effluent in a 2-reactor system (see further) can be produced by selection of the appropriate temperature, pH, substrate availability and ammonia and nitrous acid inhibition level in order to washout nitrite oxidizers from the system. In view of the Anammox stoichiometry, care should further be taken that only half of the ammonium is oxidized and that the partial nitritation is well controlled [26] as Anammox organisms are sensitive to substrate shocks.

Further, it should also be noted that due to adaptation of the biomass, nitrate build-up is possible, even after long-term operation [27].

Free Ammonia (NH$_3$) and Free Nitrous Acid (HNO$_2$) Concentration

Free ammonia (NH$_3$) and free nitrous acid (HNO$_2$) concentration have a large influence as these uncharged nitrogen forms are the actual substrate/inhibitor for ammonium and nitrite oxidation instead of ammonium (NH$_4^+$) and nitrite (NO$_2^-$) [28] and [29]. This was clearly confirmed by Van Hulle et al. [30] for ammonium oxidizers active in a SHARON reactor [27].

From the regular total ammonium (TAN) and total nitrite (TNO$_2$) analysis which gives the sum of the ionized and unionized compounds, the free ammonia (NH$_3$) and nitrous acid (HNO$_2$) concentration can be calculated incorporating pH and temperature (°C) [29]:

$$[NH_3] = \frac{[TAN]10^{pH}}{e^{6344/(T+273)} + 10^{pH}} \tag{9}$$

$$[HNO_2] = \frac{[TNO_2]10^{-pH}}{e^{-2300/(T+273)} + 10^{-pH}} \tag{10}$$

The ratio between the charged (NH$_4^+$ and NO$_2^-$) and the uncharged form (NH$_3$ and HNO$_2$) is determined by the pH and temperature values in the reactor and can be calculated based on the acid–base equilibrium. The amount of ammonia increases with increasing pH, while the amount of nitrous acid decreases which obviously promotes ammonium oxidizers but suppresses nitrite oxidizers. Hence, nitrite oxidizers can be outcompeted in a weak alkaline environment (7.5–8) in order to produce an Anammox-suited effluent in the nitritation reactor. However, the potential of using this engineering approach seems somewhat limited since adaptation of the nitrite oxidizing bacteria has been reported [31]. Therefore, the achievement of stable partial nitritation will only occur when factors other than free ammonia and free nitrous acid are regulated [25].

Concerning inhibition it can be stated that NH$_3$ is the main inhibitor of nitrification at high pH (>8), whereas HNO$_2$ is the main inhibitor at low pH (<7.5). In literature different threshold values were proposed for nitrification inhibition [29] and [32], but these are very sensitive to bacteria adaptation. Anthonisen et al. [29] stated that aerobic ammonia oxidizers are inhibited at NH_3 concentrations of 8–120 mg N L^{-1} and HNO_2 concentrations of 0.2–2.8 mg N L^{-1} while inhibition of nitrite oxidation is observed at an NH_3 concentration of 0.08–0.82 mg N L^{-1} and a HNO_2 concentration of 0.06–0.83 mg N L^{-1}. Recently, Hawkins et al. [33] stated that free ammonia has only a limited impact on the inhibition of nitrite oxidation. They found that pH changes and ammonia oxidizing activity had a bigger influence on nitrite oxidizing activity.

pH

Despite a wide divergence of the reported effects of pH on nitrification, there seems to be a consensus that the optimum pH for both ammonium oxidizers and nitrite oxidizers lies between 7 and 8. A first explanation is the influence of pH on the NH_4^+/NH_3^{-1} and HNO_2/NO_2^{-1} equilibria. The preference of ammonium oxidizers for slightly alkaline environments probably is the fact that these organisms use NH_3 as substrate [28] while at certain pH values NH_3 and HNO_2 can exhibit inhibitory effects as stated above. Apart from the influence of pH on chemical equilibria in which the substrate/inhibitors are involved, direct pH effects on the activity exist [30]. Hellinga et al. [34] observed a decrease by 8 in the growth rate of nitrite oxidizers at pH 7 compared with pH 8 whereas the variation in growth rate of ammonium oxidizers at these pH values is negligible. Below pH 7, nitrification rate will decrease since carbon limitation due to CO_2 stripping will occur [35] and [36]. However, high nitrification rates at low pH were detected in a fluidized bed reactor with chalk as biofilm carrier [37]. In this system the chalk probably acted as a local buffer system.

Other Substrates than Nitrogen

As stated above, 2 mol of protons are produced for every mol ammonium oxidized. To neutralize the acidifying step, 2 mol bicarbonate are needed. In order to assure that only half of the ammonia is oxidized to nitrite 1 mol of base per mol ammonium is required [5] and [38]. Guisasola et al. [36] and Wett and Rauch [35] reported a reduction in ammonia oxidizing activity due to bicarbonate limitation. Moreover, Ganigué et al. [39] showed that bicarbonate is a key parameter for controlling the ammonium to nitrite molar ratio in the effluent. Long-term stable nitrite build-up in a SBR treating raw urban landfill leachate was possible with extremely high ammonium concentrations (NLR = 1.2 kg N m^{-3} d^{-1}) by controlling the bicarbonate concentration. Sludge reject water is a good waste stream for partial nitritation since a proper ammonium: alkalinity ratio of 1 is often found in these streams [5]. Nitrite oxidation might also be affected by phosphorus deficiency [40]. In a biological pre-treatment plant treating highly nitrogenous wastewaters (T > 25 °C), nitrite oxidation was substantially reduced at phosphate levels below 0.2 mg P L^{-1}. Indeed, the phosphate half-saturation coefficient for nitrite oxidizers is about one order of magnitude higher than for ammonium oxidizers (0.2 mg P L^{-1} for nitrite oxidizers and 0.03 mg P L^{-1} for ammonium oxidizers) [40]. Nitrite oxidizers are especially unable to oxidize nitrite to nitrate in the absence of phosphates, the so-called phosphate block.

Temperature

Temperature is a key parameter in the nitrification process, but the exact influence is difficult to determine because of its interaction with mass transfer, chemical equilibria and growth rate. A temperature rise creates two opposite effects: increased NH_3 inhibition and increased activity of the organisms according to the Arrhenius principle. This increased activity only holds up to a certain critical temperature above which biological activity decreases again.

Experiments with pure cultures gave an optimal temperature of 35 °C for ammonium oxidizers and 38 °C for nitrite oxidizers [41]. Van Hulle et al. [30] showed that temperatures between 35 and 45 °C are optimal for partial nitritation. However, only short-term effects were investigated. Long-term exposure to temperatures above 40 °C is expected to lead to deactivation [42]. Hellinga et al. [34] concluded that temperatures above 25 °C lead to an increase of the specific growth rate of ammonia oxidizing bacteria, which become higher than that of nitrite oxidizing bacteria. The SHARON process (Single reactor High activity Ammonia Removal over Nitrite) is based on this principle. In this process, nitrification of ammonium to nitrite is established in a chemostat by working at high temperature (above 25 °C) and maintaining an appropriate sludge retention time (SRT) of 1–1.5 days, so that ammonium oxidizers are maintained in the reactor, while nitrite oxidizers are washed out and further nitrification of nitrite to nitrate is prevented (Fig. 2). Literature values for activation energy of ammonium and nitrite oxidizers range from 72 to 60 kJ mol^{-1} and from 43 to 47 kJ mol^{-1}, respectively (determined in studies between 7 and 30 °C) [43], [44],[45] and [46] indicating that the activity of ammonium oxidizers will increase faster than the activity of nitrite oxidizers.

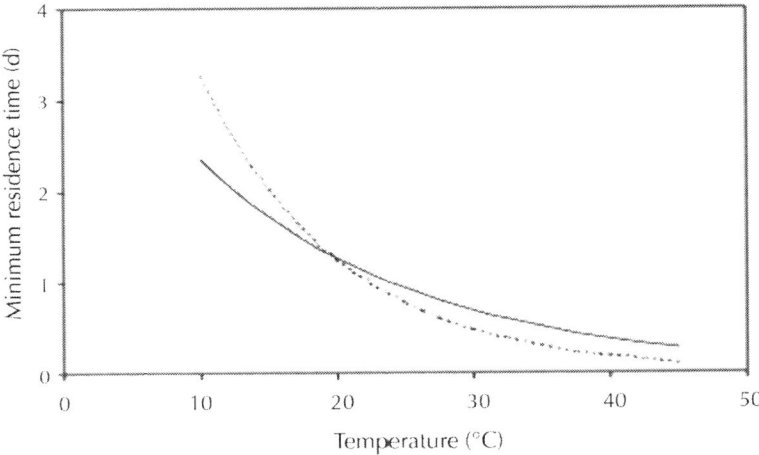

Figure 2: Effect of temperature on the minimal required cell residence time for ammonia and nitrite oxidation. Above 25 °C it is possible to wash out the nitrite oxidizers (-) while maintaining the ammonium oxidizers (· ·). Calculated with parameter values described in Jetten et al. [43].

The partial nitritation process was also successfully started up and maintained at lower temperature (between 15 and 30 °C) [47]. These results indicate that the application of the partial nitritation process could be not restricted to effluent with temperatures around 30 °C such as the effluent from methanogenic reactors but could be applicable to many kinds of industrial wastewater treatments. However, the performance of nitritation dramatically decrease below 15 °C.

Dissolved Oxygen Concentration

When it comes to nitrification, the dissolved oxygen concentration is of high importance for both ammonium oxidizers and nitrite oxidizers [48]. Ammonium oxidizers seem to be more robust against low dissolved oxygen concentration than nitrite oxidizers. Accumulation of nitrite at low dissolved oxygen is usually explained by the difference in oxygen half saturation constant (K_O) for ammonium oxidizers and nitrite oxidizers [49]. In other words, oxygen deficiency due to low dissolved oxygen concentration influences the activity of nitrite oxidizers more significantly than that of ammonium oxidizers [48]. This difference could be explained by the higher energy released per amount of oxygen consumed of ammonium oxidizers compared to nitrite oxidizers.

According to Hunik et al. [50] the half-saturation constant for dissolved oxygen is 0.16 mg O_2 L^{-1} and 0.54 for the ammonium oxidizer Nitrosomonas europaea and nitrite oxidizer Nitrobacter agilis, respectively. However, values for the half-saturation constant given in literature for activated sludge vary in the range of 0.25–0.5 mg O_2 L^{-1} and 0.34–2.5 mg O_2 L^{-1}, respectively [51]. This variation is probably due to the variation of the oxygen mass transfer efficiency in the reactors [52]. The oxygen concentration inside a sludge floc or biofilm not necessarily equals that of the water phase. The half saturation constant is therefore dependent on the biomass density, the floc size, the mixing intensity and the rate of diffusion of oxygen in the floc [53]. This was clearly demonstrated by Manser et al. [54]. They showed that the half-saturation constants for oxygen determined for sludge coming from a conventional activated sludge plant and sludge coming from a membrane bioreactor exhibited a major difference because sludge flocs in the membrane bioreactor are much smaller. Hence, diffusion resistance in these flocs can be neglected.

Peng et al. [55] and Jubany et al. [56] demonstrated that it was possible to remove nitrogen through nitrification–denitrification over nitrite by using an on-line dissolved oxygen or OUR control system. This system controls the oxygen concentration by turning aeration off at the point when ammonia oxidation had completed. This point was determined from the pH and dissolved oxygen signals. As such nitrite oxidation was prevented by limiting the oxygen supply. Aeration patterns are proposed to be an alternative parameter to control ammonium to nitrite [57]. Hyungseok et al. [58] reported that nitrate formation can effectively be prevented by frequent switching between oxic and anoxic phases. As such the aeration is switched off before all the ammonium is consumed and before nitrite can be further converted to nitrate. A prolongation of the aeration phases in a SBR lowered the stress on nitrite oxidizers resulting in an increase of nitrate [59] and [60]. These findings were confirmed by Sin et al. [61] who found that nitrite build-up was caused by low oxygen concentration (0.5 mg O_2 L^{-1}) and fast alternation of the aeration conditions in the system. Imposing oxygen-limiting conditions can be considered another way to outcompete nitrite oxidizers. However, it is also suggested that free hydroxylamine inhibition of nitrite oxidizing bacteria rather than a difference in oxygen affinity constants causes nitrite build-up in nitrifying systems at low dissolved oxygen concentration [62]. According to Hu [63] hydroxylamine exhibited acute and irreversible toxicity to Nitrobacter (nitrite oxidizers) and this may also cause nitrite build-up in a nitrifying system. Castignetti and Gunner [64] and Stüven et al. [65] stated also that hydroxylamine severely inhibits nitrite oxidizers.

Sludge Age

Ammonium oxidizers and washout of nitrite oxidizers can be selectively accomplished by the application of an appropriate sludge retention time in suspended growth systems because of different minimum required sludge ages. The minimum doubling time for ammonium oxidizers is 7–8 h and for nitrite oxidizers 10–13 h, respectively [66]. Van Kempen et al. [67] found on the basis of full-scale experience that a maintenance of SRT between 1 and 2.5 days resulted in good performance. Selection of the AOB on the basis of different growth rates are used in the SHARON process. This process operates at a HRT (equal to SRT) of 1 day under high temperature and high oxygen concentration to favour the growth of ammonia oxidizers and to washout the nitrite oxidizing bacteria. However, successful partial nitritation were reported under longer sludge age. Pollice et al. [68] and [69] showed that a stable partial nitritation could be obtained under oxygen limitation independent of the sludge age of 10, 14 and 40 days while Peng and Zhu [25] also fulfilled stable nitrite accumulation under normal or even low temperature (<13 °C) with extended SRT (30 days) (Table 2).

Table 2: COD, BOD, N and P concentrations (in mg L^{-1}) in waste streams with high nitrogen content

Type waste-water	COD	BOD$_5$	Total nitro-gen	Phospho-rus	Reference
Reject water	232–12587	81–750	260–958	33–207	[179]
	390–2720	n.m.	943–1513	n.m.	[180]
	610	140	910	n.m.	[181]
Thin frac-tion piggery manure	n.m.	2912	707	55	[182]
	3969	1730	1700	147	[183]
	9000–13000	n.m.	3100–4300	20–40	[184]
	6456	n.m.	695	91.8	[185]

Landfill leachate	2C00–5000	1500–4000	500–1000	20–50	[186]
	n.m.	45	310	n.m.	[14]
	1300–1600	n.m.	160–270	n.m.	[187]
	9660–20560	n.m.	780-1080	20–51	[188]
Tannery waste water	300–1400	n.m.	50–200	n.m.	[22]
	1940–2700	n.m	123–185	n.m	[17]
Slaughter house waste processing	1400–2400		170–200	35–55	[18][a]
Starch pro-duction	3000	990	1060	210	[189][b]
	50C0–10000	2000–5000	800–1100	170–230	[19][b]
Pectine in-dustry waste water	15000–22000	n.m	1280–2990	n.m	[190]
	8100	n.m.	1600	11	[191]

n.m.: not mentioned.

[a] After treatment in anaerobic lagune.

[b] After treatment in anaerobic digester.

Organic Carbon and Salts

It is stated that the partial nitritation process is successful for the treatment of wastewater with low C N^{-1} ratio although other streams with high organic content and high ammonium concentration such as swine wastewater [70] and [71] and monosodium-glutamate manufacturing streams are also used in partial nitritation processes [72].

Mosquera-Corral et al. [73] observed stimulation of the ammonia oxidation in the SHARON-process when acetate as carbon source was fed in a 0.2 g C g N^{-1} ratio leading to an effluent with nitrite to ammonia molar ratios higher than the stoichiometric ones. On the other hand, an inhibitory effect of ammonia oxidizing activity of 10% was observed when 0.3 g C g N^{-1} was brought into the reactor. Hanaki et al. [49] suggested that this inhibition was caused by a decreasing affinity of ammonia oxidizers for ammonia. One possible explanation is that the transport of ammonia from the bulk water phase to the cell of the ammonia oxidizer could be hindered by the presence of the crowded cells of heterotrophs which assimilate the ammonia and consume the oxygen before it reaches the nitrifiers. However, Hanaki et al. [49] found that for the same SRT, the ammonia oxidation efficiency decreased at higher COD concentrations but at a constant COD concentration efficiency restored again by increasing the SRT. Therefore, a moderate increase of the SRT to 2–3 days could be a possible solution to minimize the effect of heterotrophs on the ammonia oxidation.

Many industrial wastewaters rich in ammonium also contain high salt concentrations which could inhibit ammonia oxidation. However, the SHARON process still occured successfully at high NaCl concentrations of 100 mM in case of batch experiments and 427 mM in continuous operation [73]. This different behaviour was attributed to the adaptation of biomass to the saline environments.

Other Influencing Factors

Research by Zepeda et al. [74] showed that benzene, toluene and xylene induce a significant decrease in the values for nitrification specific rates affecting mainly the ammonia oxidation pathway. Heavy metals chromium, nickel, copper, zinc, lead and cadmium might inhibit both steps of nitrification reaction but the inhibition effects are different (Table 3; [75]).

Table 3: Inhibition to nitrification of some metal concentrations under pure culture [75]

Metal	Concentration (μg L^{-1})
Cr	0.7–785
Ni	3–860
Cu	3–5730

Zn	3–1000
Pb	0.09–1680
Cd	0.01–20

Formic, acetic, propionic and n-butyric acid all inhibited nitrite oxidation, but exhibited no significant effect on ammonium oxidation [76].

Of a dozen compounds tested by Tomlinson et al. [77], only chlorate, which is used to stop nitrite oxidation [78], cyanide, azide and hydrazine were inhibitorier to the oxidation of nitrite than to the oxidation of ammonium. Other toxic components that influence nitrite oxidation are the disinfectants bromide and chloride [79].

Light is inhibiting both ammonium oxidizers and nitrite oxidizers since cytochrome c is oxidized by light in the presence of oxygen.

THE ANAMMOX PROCESS

When Mulder et al. [3] observed unexplainable nitrogen losses in denitrifying fluidized bed reactors the idea was put forward that this could be attributed to ANaerobic AMMonium Oxidation (Anammox). Twenty years before, Broda [4] predicted that this process was possible on the basis of thermodynamic calculations. Van de Graaf et al. [80] showed by inhibition experiments that Anammox is a microbially mediated process and not a chemical reaction. ^{15}N labelling experiments showed that nitrite was the preferred electron acceptor instead of nitrate as first assumed [81]. Hydroxylamine and hydrazine were identified as important intermediates [43]. By making the mass balances over different Anammox enrichment cultures the overall stoichiometry of the Anammox reaction was determined as expressed in equation (11) [82]

$$NH_4^+ + 1.32NO_2^- + 0.066HCO_3^- + 0.13H^+ \rightarrow 1.02N_2 + 0.26NO_3^- + 0.066CH_2O_{0.5}N_{0.15} + 2.03H_2O \quad (11)$$

The Anammox process involves the oxidation of ammonia into dinitrogen gas in absence of oxygen [82]. This implies that the name anaerobic ammonium oxidation should actually be anoxic ammonium oxidation since nitrite is present as electron acceptor. It was found that nitrite was not only used for the oxidation of ammonium, but it was also oxidized to nitrate. This oxidation generates the reducing equivalents necessary for carbon fixation [81] and [83]. Since Anammox bacteria are autotrophic, the conversion of ammonia into dinitrogen gas can take place without addition of organic matter [84].

Since the initial discovery of Anammox activity has been reported in different wastewater treatment facilities [85], ranging from installations treating wastewater with high nitrogen load at low dissolved oxygen concentrations [86] to municipal wastewater treatment plants [87]. Further, Anammox is present in different natural environments and contributes significantly to the world's nitrogen cycle as it is found in several of the world's seas and rivers such as the Black Sea [88] and the Thames estuary [89]. Depending on the organic load up to 70% of the N_2 production in marine sediments can be attributed to Anammox [90]. Strous et al. [91] showed that the bacteria responsible for the Anammox process are new members of the order of the Planctomycetes (Table 4). Fluorescence in situ hybridisation (FISH) probes were developed for the different Anammox species. Based on FISH analysis Schmid et al. [92] concluded that different Anammox species rarely occur in the same WWTP or enrichment culture. It seems that they all occupy their own niche and environmental conditions select for only one of the different genera [92]. However, Furukawa et al. [93] found two different genera of Anammox bacteria in a lab-scale reactor in which partial nitritation and the Anammox process are established together. Important work was performed concerning Anammox phylogeny [92], Anammox biochemistry [81], [94], [95] and [96], Anammox compartmentalization [97] and [98] and Anammox unconventional membrane lipids [99].

Table 4: Biodiversity of Anammox bacterial species

Genus	Species	Source	Reference
Brocadia	Candidatus Brocadia anammoxidans	Wastewater	[91]
	Candidatus Brocadia fulgida	Wastewater	[192]
Kuenenia	Candidatus Kuenenia stuttgartiensis	Wastewater	[193]
Scalindua	Candidatus Scalindua brodae	Wastewater	[92]
	Candidatus Scalindua wagneri	Wastewater	[92]
	Candidatus Scalindua sorokinii	Seawater	[88]
	Candidatus Scalindua Arabica	Seawater	[194]
Others	Candidatus Jettenia asiatca	Not reported	[121]
	Candidatus Anammoxoglobus propionicus	Wastewater	[120]

Anammox biomass has a brown-reddish colour, which is probably due to the high cytochrome contents [43]. In last decade several techniques were developed for the detection of active Anammox organisms [85]. In Table 4 the different Anammox species are presented.

According to Schmid et al. [92], Anammox organisms have doubling time of 11 days and a biomass yield of 0.13 g dry weight per g NH_3-N oxidized. However, van der Star [100] concluded that the doubling time of Anammox bacteria is at most 5.5–7.5 days calculated on the basis of maximum conversion capacity, but possibly as low as 3 days. Researchers recently have claimed they optimized the reactor conditions to such an extent that a doubling time of 1.8 days was achieved [101]. A possible explanation for this high variation in growth rate could be the method to determine the growth rate, as Isaka et al. [101] determined the growth rate by direct counts of Anammox bacteria, while other studies rely on biomass yield and nitrogen removal rate.

This low growth rate and the difficulty in obtaining axenic cultures strongly hindered Anammox research [82] and [91].

Inhibition of Substrates and Products

The nitrite concentration is an important parameter to control since Anammox activity is inhibited by it. However, no uniformity is found about the threshold values of nitrite inhibition. Dapena-Mora et al. [102] found that 350 mg N L^{-1} nitrite correspond to 50% inhibition of the Anammox process performing activity tests. In the presence of more than 100 mg N L^{-1} nitrite, Strous et al. [103] found that the Anammox process was completely inhibited. Fux [104] showed in a long-term experiment that maintaining a nitrite concentration of 40 mg N L^{-1} over several days led to the irreversible inactivation of the Anammox organisms. This decreased activity due to nitrite inhibition can be restored by adding trace amounts of the Anammox intermediates hydroxylamine and hydrazine, even after long-term exposure to high concentrations of nitrite [103].

Remarkable is also the difference in tolerance for nitrite between the different Anammox genera. The inhibition experiments conducted by Strous et al. [103] were performed with Candidatus Brocadia anammoxidans. Experiments of Egli et al. [105] with Candidatus Kuenenia stuttgartiensis showed that the Anammox process was only inhibited at nitrite concentrations higher than 182 mg N L^{-1}. Furthermore, experiments by Strous et al. [103] showed that increasing the nitrite concentration changed the stoichiometry of ammonium and nitrite consumption from 1.3 g nitrite per gram ammonium at 0.14 g N L^{-1} nitrite to almost 4 g nitrite per gram ammonium at 0.7 g N L^{-1} nitrite. From the distorted stoichiometry at high nitrite concentrations, it can be concluded that the microorganisms under these conditions did not only use ammonium as the electron donor but also must have generated an internal electron donor to reduce the nitrite. This changing stoichiometry was also noticed at higher temperatures. Dosta et al. [106] observed an nitrite:ammonium consumption ratio of 1.38:1 at a working temperature of 30 °C but this ratio decreased to 1.05:1 when the reactor was operated at 18 °C.

The Anammox process is not inhibited by ammonium or by the by-product nitrate up to concentrations of at least 1 g N L^{-1} [103].

Dapena-Mora et al. [102] observed a 50% activity loss with high concentrations of ammonium and nitrate (770 and 630 mg N L^{-1}, respectively).

It is known that the chemolithoautotrophs mainly utilize inorganic carbon as carbon source. Therefore, the influent bicarbonate concentration is an important factor to affect the Anammox enrichment. Dexiang et al. [107] observed low Anammox activity at low bicarbonate: ammonium ratios of 2.3:1. At these conditions a limitation of the activity could occur since not enough CO_2 is present. On the other hand, high bicarbonate concentrations (bicarbonate:ammonium ratio of 4.7:1) lead also to inhibition. A possible explanation could be the formation of a high amount of free ammonia since the pH in reactor reached 8.1.

Phosphate and Sulphide

Similarly to nitrite inhibition a difference in tolerance for phosphate exists between different Anammox species. van de Graaf et al. [83] experienced a loss of activity for C. Brocadia anammoxidans at phosphate concentrations above 155 mg P L^{-1}, while Egli et al. [105] did not see any inhibitory effect of phosphate when a culture of C. Kuenenia stuttgartiensis was supplied with up to 620 mg P L^{-1}. Dapena-Mora et al. [102] observed at the same phosphate level of 620 mg P L^{-1} 50% inhibition of Anammox activity. In batch tests using sludge from a highly loaded lab-scale rotating biological contactor containing C. Kuenenia stuttgartiensis, phosphate was shown to partially inhibit the Anammox process [108]. Anammox activity decreased to 63% of the normal activity at 55 mg P L^{-1} and further to 20% at 110 mg P L^{-1}. At 285 mg P L^{-1} no further decrease was observed (80% inhibition).

The effect of sulphide on the activity was also tested since SO_4^{2-} reduction often takes place in anaerobic digestion mainly transformed into H_2S. In anaerobic conditions, sulphate reducing bacteria produce sulphide with organic carbon as electron donor. Wastewaters such as seafood processing, leather tanning, and oil refining and alcohol fermentation not only contain organic carbon and nitrogen but also sulphur compounds. Dapena-Mora et al. 2007 [102] showed an Anammox inhibition of 50% at low sulphide concentration of 9.6 mg S L^{-1} while van de Graaf et al. [83] showed a resistance of Anammox to at least 64 mg S L^{-1} in continuous and batch experiments. This large difference in sulphide inhibition could be explained by the addition of nitrate as electron donor for the Anammox biomass in van de Graaf et al. [83] since sulphide could reduce nitrate to nitrite, which is the preferable electron donor of the process. Recently, simultaneous removal of ammonium and sulphate by Anammox have been reported by Yang et al. [109].

Oxygen

Anammox bacteria are strictly anaerobic and are inhibited by dissolved oxygen. Inhibition caused by low concentration of oxygen was demonstrated however to be reversible. Egli et al. [105] stated that oxygen inhibits Anammox metabolism reversibly at low oxygen levels (air saturation of 0.25–2%) but probably irreversibly at high levels (>18% air saturation). Strous et al. [110] concluded from experiments with intermittent oxygen supply that the Anammox process was reversible inhibited by oxygen, making partial nitritation and Anammox possible in one reactor [110].

Organic Carbon

Landfill leachate and wastewaters from digested animal waste contain high nitrogen concentration but also high organic carbon levels. Still, there are considered to be good influent streams for Anammox reactor. During anaerobic digestion fast biodegradable organic content is converted to biogas. As such, only slow biodegradable organic matter will be present in these wastewaters. Ruscalleda et al. [111] found that Anammox and denitrifiers could co-exist and play an important role in treating streams with high quantities of slowly biodegradable organic carbon such as digested liquor and landfill leachate. In such streams, heterotrophic denitrifying growth is limited by the low availability of easily biodegradable organic carbon. As consequence denitrifiers are not able to dominate in these systems and could not outcompete Anammox organisms. Undigested animal streams contain high nitrogen concentration but also high organic carbon levels. Since most of the fast degradable organic carbon content is oxidized in the proceeding partial nitritation step, the content of organics would be low enough so that denitrifiers does not outcompete Anammox.

Several other studies reported that presence of organic matter has a negative impact on Anammox growth [43], [71], [112], [113], [114], [115] and [116]. In presence of certain amounts of organic carbon, Anammox organisms are no longer able to compete for nitrite with denitrifiers. This could be due to the fact that the growth rate of denitrifiers is higher than Anammox bacteria [91]. Moreover, the denitrification reaction is thermodynamically more favourable than anaerobic ammonium oxidation (the Gibbs free energy of Anammox bacteria is -355 kJmol^{-1} [43]), while the Gibbs free energy of denitrifying bacteria is -427 kJmol^{-1} [117]). Therefore, heterotrophic denitrifiers would grow faster when organic carbon is present in combination with ammonium and nitrite eliminating place for Anammox organisms. The threshold concentration for organic carbon in which denitrifiers outcompete Anammox bacteria differs from report to report. Güven et al. [115] stated that Anammox bacteria are no longer able to compete with heterotrophic denitrifying bacteria at C N^{-1} ratio above 1 while Chamchoi et al. [114] stated that an organic matter concentration above 300 mg COD L^{-1} or COD to N ratio of over 2.0 inactivate Anammox organisms in a UASB reactor fed with fat milk as organic carbon source. Milenuevo et al. [71] observed a complete inhibition of the Anammox process at COD concentrations up to 292 mg L^{-1} while Tang et al. [113] stated that denitrifiers became dominant at high influent COD:NO$_2^-$ N ratio of 2.9:1.

Anammox removes only 90% of the incoming nitrogen as ammonia/nitrite and leaves 10% of nitrogen as nitrate in the effluent. A co-existence of Anammox and denitrification in one reactor would aid to reduce the nitrate concentration in the reactor.

Under anoxic conditions nitrate can be reduced by denitrifiers to nitrite as intermediate which can be utilized by Anammox for the oxidation of ammonium [118].

Anammox activity is completely and irreversible inhibited by low concentrations of methanol (15 mg L^{-1}) and ethanol [115]. This aspect must be taken in account since methanol is often used to remove nitrate in a post-denitrification step. A possible explanation for the methanol inhibition is the formation of formaldehyde by the Anammox enzyme hydroxylamine oxidoreductase [119].

Recent studies observed that some organic carbon sources do not have an inhibition effect on the Anammox acitivity. Kartal et al. [120] reported that Candidatus Brocadia fulgida and CandidatusAnammoxoglobus propionicus are able to oxidize acetate and propionate, respectively. Experiments by Güven et al. [115] with propionate as carbon source showed that Anammox organisms oxidized propionate with nitrate and/or nitrite as electron acceptor and simultaneously converted ammonia anoxically. The amount of Anammox bacteria and denitrifiers did not change over time, suggesting that Anammox organisms are indeed able to compete with heterotrophic denitrifiers for propionate. Awata et al. [122] used batch test to investigate the ammonium removal and the carbon incorporation by the Anammox bacteria in the presence of short chain fatty acids present in digestor liquor such as acetate, formate and propionate. They found that propionate did not influence the ammonium removal activity but decreased the incorporation of inorganic carbon. Acetate showed no inhibition in ammonium removal and inorganic carbon incorporation while formate inhibited the Anammox process in the two aspects. It is not yet known whether the Anammox bacteria incorporate acetate directly or indirectly. It could be possible that the CO_2 used by Anammox was derived from denitrification with organic matter such as acetate. Experiments with Anammox cultures in batch experiments by van de Graaf et al. [83] showed that carbon sources such as acetate and glucose had a positive effect on Anammox activity. The continuous experiments however, fed with acetate, glucose and formate showed a negative effect on Anammox activity

[83]. Dapena-Mora et al. [102] used batch tests to observe the effect of inhibition effects on the Anammox performance. They found that concentrations of 50 mM acetate resulted in 70% inhibition of the Anammox process while a concentration up to 10 mM did not decrease the activity significantly [102].

Adaptation of Anammox bacteria to streams with toxic components are reported. Toh and Osbolt [123] and Toh et al. [124] described an acclimation of the Anammox organisms to synthetic coke-oven wastewater which contain not only high concentration of organics (2000–2500 mg L^{-1} COD) but also some toxic chemicals such as phenol (300–800 mg L^{-1}), cyanides (10–90 mg L^{-1}) and thiocyanates (300–500 mg L^{-1}). The initial attempt to enrich the bacteria first failed but stepwise addition of phenol 50–500 mg L^{-1} aided to acclimate the Anammox. After a culture of 15 months, the ammonium removal rate peaked to 0.062 kg N m^{-3} d^{-1}.

Salts

In natural saline ecosystems only anammox species belonging to Scalindua genus have been detected [90]. The other genera are known to inhabit freshwater ecosystems [125].

Dapena-Mora et al. [102] found that NaCl concentrations below 150 mM did not affect the Anammox activity while KCl and Na$_2$SO$_4$ had affect at concentrations higher than 100 and 50 mM, respectively. They stated that the different inhibitory effect of NaCl and Na$_2$SO$_4$ was attributed to the ion sodium at the tested concentration. Hence, the inhibitory effect of Na$_2$SO$_4$ could be related to the higher concentration of Na$^+$ions in the medium compared to its concentration when NaCl was added at the same molarities. Van de Graaf et al. [83] observed no effect of KCl on the activity at concentrations of 50 mM. Kartal et al. [125] reported the adaptability of a freshwater Anammox biomass, i.e. C. Kuenenia stuttgartiensis, to salt concentrations as high as 30 g L^{-1} in a lab-scale investigation provided that salt concentrations was gradually increased. The nitrogen removal efficiency and maximum Anammox activity of the salt adapted sludge was very similar to the reference freshwater sludge. Windey et al. [126] operated an OLAND under saline conditions and came to the same conclusions as Kartal et al. [125].

Temperature and pH

Several authors found that the optimum temperature for the operation of Anammox bacteria was around 30–40 °C [103] and [105]. Dosta et al. [106] used batch tests to observe the short-term effect of temperature on Anammox activity. They found that the maximum activity of non-adapted Anammox biomass ranged between 35 and 40 °C, while a temperature of 45 °C caused an irreversible decrease of the Anammox activity due to biomass lysis. Small differences in optimal temperature were found for K. stuttgartiensis and B. anammoxidans. B. anammoxidans showed highest activity at 40 °C [103] while the highest activity of K. stuttgartiensis was observed at 37 °C [105] at an optimal pH of 8.

However, Cema et al. [127] and Isaka et al. [128] proved that the Anammox process in a RBC and anaerobic biological filtrated reactor, respectively could be successfully operated at a low temperature of 20 °C. The slow adaptation of the Anammox sludge seems a key factor in order to operate an Anammox reactor at low temperatures since a drastic change in the operational conditions could lead to a destabilization of the biological system [129]. During the operation of this reactor at low temperature, neither changes on the physical properties of sludge nor qualitative changes on the bacterial populations were found. However, a strong decrease in the nitrogen conversion rate was observed. Isaka et al. [128] solved this problem and reached a high nitrogen conversion rate of 8.1 kg N m^{-3} d^{-1} by decreasing HRT and by using an appropriate non-inhibiting nitrite concentration in the influent.

As stated above, an advisable start up strategy is needed to operate a Anammox system at low temperatures. First, the required amount of biomass must be produced in a separate reactor at a temperature close to the optimum temperature. Then, the biomass can be gradually adapted to low temperatures in the same reactor and finally the low-temperature adapted biomass can be inoculated in the low-temperature reactor [106]. Moreover, research performed with marine Anammox samples reported measurable activities at low temperatures. Rysgaard et al. [130] found Anammox activity in Arctic sediments between −1.3 and 30 °C, the optimum temperature being 12 °C. Similar results were found by Dalsgaard and Thamdrup [90] giving an optimal temperature of 15 °C for marine sediments from the Baltic North Sea. In both cases, a strong decrease of the nitrogen conversion rate was observed.

The optimal pH interval for Anammox is 6.7–8.3 with an optimum of 8.0 [103].

Biomass Concentration

Biomass concentration plays a crucial role for the Anammox activity. Strous et al. [91] found that Anammox is only active when cell concentrations are higher than 10^{10}-10^{11} cells ml^{-1}, even in purified cultures. This could be explained by the need for intercellular communication for activity [34]. Another possible explanation is that hydrazine diffuses relatively easy to the outside of the cell and a minimum internal concentration is necessary for Anammox activity. Sinninghe Damsté et al. [99] however showed that the cellular membranes are less permeable than normal linear membrane lipids.

Perhaps the presence of contaminating cells, 1 on 200–500, is necessary to sustain long term growth, because these cells can guarantee vitamin supply and the removal of toxic components [91] and [95]. Pynaert et al. [131] put forward the hypothesis that the presence of ammonium oxidizers is necessary for the re-activation of Anammox organisms after disturbance of the system. By the production or accumulation of hydroxylamine or hydrazine by ammonium oxidizers, Anammox organisms can re-activate their metabolism. Once the process is re-established, the ammonium oxidizers are not supposed to significantly participate in the Anammox process. This "sparking" was also described by Strous [132] because it was found that the addition of the intermediates hydroxylamine or hydrazine was necessary to restart Anammox activity after inhibition. On the other hand, Dapena-Mora et al. [102] did not observe a notable effect on the activity at different initial biomass concentration of $0.25–2.0$ g VSS L^{-1}.

Suspended Solids

Flocculants are often used to remove colloidal organic and inorganic substances from wastewater previous to the Anammox process. Therefore, the effect of these flocculants on the Anammox process are tested in batch tests by Dapena-Mora et al. [102]. Concentrations up to 1 g L^{-1} polymeric positively charged compound used as flocculant did not cause a detrimental effect on the Anammox activity. In the study of Yamamoto et al. [133], a large amount of influent suspended solids present in the partial nitrified digested liquor attached to the nonwoven materials covering the anammox biomass growing on the carries. This caused a decrease in Anammox activity and became the main reason responsible for the unsatisfactory performance of the Anammox process. The use of a flocculant improved the settle ability of the influent suspended solids and reduced their accumulation inside the reactor but the flocculant itself attached also on the surface of the nonwoven carriers and hence reducing Anammox activity. An unstable operation of the Anammox reactor is also reported due to precipitation of phosphate salts. Trigo et al. [134] operated an Anammox MSBR in which the membrane acted as a barrier that retained the inorganic precipitation salts causing an accumulation of non-suspended volatile solids in the biomass. Precipitation of these salts on the biomass surface interfered with microbial activity causing a decrease of nitrogen removal from 100 to 10 mg L^{-1} per day.

Other Influencing Factors

Anammox activity was also found to be sensitive to visible light. A decrease in activity of 30 to 50% was observed by van de Graaf et al. [83]. As a result the equipment for further experiments by these researchers was covered with black plastic and paper to eliminate this light effect. Arrojo et al. [135] showed the effect of shear stress on the Anammox process in a SBR. They stated that stirring speeds up to 180 rpm had no negative effect on the performance of the Anammox process. Anammox activity and the average diameter decreased to 40% and 45%, respectively while nitrite accumulated in the reactor when a rotating speed of 250 rpm was tested.

PRACTICAL IMPLEMENTATION OF AUTOTROPHIC NITROGEN REMOVAL

An Anammox step has to be preceded by a partial nitritation step. This can be accomplished in the same reactor (1-reactor system) or by using 2 separate reactors (2-reactor system).

The use of a single reactor has some advantages with respect to the partial nitritation–Anammox configuration. Single-stage processes generally have higher volumetric nitrogen removal rate and lower capital costs than 2-stage systems since no additional nitritation reactor volume is required for ammonium oxidation without nitrogen removal [136]. However, Hao et al. [137] and Nielsen et al. [138] reported difficulties in dissolved oxygen regulation and incomplete nitrogen removal treating high loaded wastewaters. With a 2-reactor system nitrititation and Anammox are separated in space allowing flexibility and a more stable process performance since both steps can be controlled separately [136] and [139]. In a first reactor half of the ammonium is converted to nitrite, while in a second reactor Anammox is active. It is important that the influent of the Anammox reactor has a constant composition in view of the nitrite toxicity, independent of the strategy used to obtain this Anammox-suited influent. The application of the two unit's configuration would be appropriated when toxic or organic biodegradable compounds are present since these compounds will be degraded in the proceeding nitritation step avoiding its entrance to the Anammox reactor [140] and [141]. Hao et al. [137] and Nielsen et al. [138] stated that for high loaded waste streams the relatively high investment costs for a partial nitritation–Anammox process will be compensated by lower operational costs and efficient nitrogen removal performance.

A major disadvantage of these autotrophic nitrogen removal processes is the low growth rate of AOB and Anammox bacteria. The performance of reactors involving slow growing bacteria can be enhanced by applying high sludge retention time. This could be achieved by applying carrier materials to develop biofilms or by self-aggregation in granules [206].

An overview of different reactors described in literature is presented here. From this overview it becomes clear that a lot of the experimental knowledge on autotrophic nitrogen removal are described in lab scale reactors although in recent years full-scale reactors are brought into operation.

Partial Nitritation and Anammox in One Reactor (1-reactor System)

In a 1-reactor system, a co-culture of aerobic and anaerobic ammonium-oxidizing bacteria is established under microaerobic conditions to avoid inhibition of Anammox bacteria by oxygen and to achieve appropriated conditions to obtain partial nitritation [110]. In those system, the growth of NOB (and subsequent nitrate production) is prevented due to their lower affinity to oxygen compared to AOB and for nitrite compared to Anammox bacteria [49]. Possible inhibition of nitrite oxidizers by free ammonium has also been suggested [19].

Different kind of systems such as SBR, gas-lift, RBC and moving bed reactors were used to obtain the microaerobic conditions for the 1-step process. In biofilm or granule reactors the ammonium oxidizers are active in the outer layers of the biofilm (or granule), producing a suitable amount of nitrite for the Anammox organisms that are active in the inner layers. This way the Anammox organisms are protected from oxygen, which is consumed in the outer layers [136]. A variation on the classic biofilm reactor is the membrane aerated biofilm reactor (MABR [142]). In MABR systems hydrophobic, gas-permeable membranes are used for bubbleless oxygen transfer. In the oxygen rich region near the membranes ammonium oxidizers are converting ammonium to nitrite, while in the ammonium rich region near the water phase Anammox organisms are active.

When these biofilms and granular systems are used to perform the process, mass transfer resistance uses to be the limiting step. As long as ammonium concentration outside the biofilm is much higher than the oxygen or nitrite concentration, ammonium diffusion into the biofilm will not limit the process rate. If the nitrite produced in the outer layer is mainly consumed in the inner layer, oxygen is the main limiting factor controlling the overall rate. Sliekers et al. [143] and Szatkowska et al. [144] reported that oxygen transfer was indicated as the limiting factor for a lab-scale air-lift and a pilot-scale moving bed reactor, respectively. This oxygen limitation can be attributed to the slow diffusion into the biofilm/granule or from a not-efficient gas–liquid transfer.

Nitrite concentration is a strong inhibitor of the Anammox process. If nitrite is consumed at about the same rate as it is produced this inhibition effect is not of significance. No negative effect of nitrite was observed by Vazquez-Padin et al. [140] although during the first 100 days of operation a mean nitrite concentration of 25 mg N L^{-1} was registered. Probably a concentration gradients inside the granules resulted in a low nitrite concentration at the location of the Anammox bacteria.

Various names are used to describe the 1-reactor systems [104]: the OLAND-process (Oxygen Limited Autotrophic Nitrification and Denitrification) [145], the CANON process (Completely Autotrophic Nitrogen removal Over Nitrite) [146], aerobic/anoxic deammonification or DEMON [147] and [148] and the SNAP process (Single-stage Nitrogen removal using Anammox and Partial nitritation) [93]. The difference lies in the organisms that were originally assumed to be responsible for anaerobic ammonium oxidation. In both the OLAND-process and the aerobic/anoxic deammonification process nitrifiers were assumed to perform this ammonium oxidation under microaerobic conditions [145] and [149]. In the CANON process Anammox bacteria were assumed to be responsible. Studies [108] and [150] with FISH analyses confirmed that anaerobic ammonium oxidation in all reactors was performed by Anammox organisms, although Pynaert et al. [108] did not exclude a specific role for the aerobic ammonium oxidizers.

Most CANON systems reported in literature were operated at 30–35 °C with a maximal nitrogen removal rate of 0.075–1.5 kg N m^{-3} d^{-1}[143] and [151]. At these temperatures AOB grow faster than NOB and also the growth of Anammox bacteria is stimulated since this temperature range lies close to their optimal temperature. However, in an air pulsating SBR operated at 20–24 °C a similar maximal nitrogen removal of 0.5 kg N m^{-3} d^{-1} [140] are reported while only a slightly activity of NOB was observed. The feasibility of achieving a quick start-up and high nitrogen removal rates in autrophic nitrogen removing systems at temperature around 20 °C was already reported by Isaka et al. [128] and Dosta et al. [106] in a two stage system and Pynaert et al. [131] in one stage system.

The efficient retention of biomass in a SBR makes it possible to cultivate slowly growing bacteria. However, higher nitrogen removal rates were obtained in a RBC [108] and [131] and in an air lift reactor [143].

Model simulations indicated that the maximum nitrogen removal rate was achieved only when the dissolved oxygen concentration kept pace with the ammonium surface load [137]. For fluctuating ammonium loading rates in engineering dissolved oxygen can be regulated through online feed-back control [138]. With a simulation study Hao et al. [137] showed that the optimal bulk oxygen concentration for a CANON biofilm reactor is about 1 mg O$_2$ L^{-1}, although this optimum depends on the biofilm thickness and density, boundary layer thickness, the COD content of the influent and the temperature. Oxygen control of these CANON systems is therefore necessary.

Kuai and Verstraete [145] first introduced the term OLAND describing lab-scale research with a SBR reactor fed with synthetic influent in which only 0.050 kg N m^{-3} d^{-1} ammonium was removed. Sliekers et al. [151] conducted experiments in lab-scale completely mixed reactors using a specific start-up pattern consisting of anoxic inoculation with Anammox biomass followed by oxygen supply to develop nitrifying microorganisms. Ammonia was mostly converted to nitrogen gas (85%) while the remainder was recovered as nitrate. However, the nitrogen removal rate in the SBR [151] was still low with only 0.064 kg N m^{-3} d^{-1}. FISH analysis confirmed the absence of nitrite oxidizers and the presence of aerobic ammonia oxidizers (45%) and anaerobic ammonium oxidizers (40%) in the CANON biomass [151]. Recently, De Clippeleir et al. [152] and Vazquez Padin [140] and [206] observed high nitrogen removal rates in a SBR provided that granulation occurred. The operation of these granular sludge reactors is very similar to biofilm reactors. In the outer oxic layer ammonium is converted to nitrite, while Anammox is active in the centre of the granule.

Pynaert et al. [108] and [131] constructed, operated and characterized an OLAND RBC system fed with both synthetic and actual waste water and a fixed oxygen concentration in which high removal rates could be achieved. Within 100 days after inoculation of a granular anaerobic sludge a maximum ammonium removal of 1.80 kg N m^{-3} d^{-1} was achieved. In Sliekers et al. [143] a gas lift reactor with high conversion rate of up to 1.5 kg N m^{-3} d^{-1} was easily maintained. Recently also artificial wetlands were used as autotrophic nitrogen removing systems resulting in 50–60% nitrogen removal [153] and [154].

The term aerobic/anoxic deammonification or DEMON was first used when significant losses of inorganic nitrogen of up to 90% were observed in the nitrification step of a rotating biological contactor (RBC) treating ammonium-rich landfill leachate under low oxygen conditions [147]. Extended nitrogen loss was also observed in other RBCs in Switzerland and the UK [86] and [155]. None of the plants were specifically built for deammonification, but nitrogen elimination was established over time. In the Swiss RBC about 50% of the bacteria population in the biofilm consisted of Anammox. Next to RBC's continuous flow moving-bed pilot plants were run as well. Optimal ammonium elimination was achieved at a bulk oxygen concentration of 0.7 mg O_2 L^{-1}. The end product is always N_2, although Gaul et al. [156] reported up to 12% N_2O production caused by incomplete heterotrophic denitrification under anoxic or oxygen-limited conditions. The first full-scale application with deliberate deammonification in a moving bed reactor using Kaldnes®carriers was put into operation in April 2001 [157] at the WWTP of Hattingen (Germany). Two identical reactors had a volume of 67 m^3 and an effective biofilm surface area of 13,400 m^2. The oxygen concentration was kept below 1 mg O_2 L^{-1}. First results are given in Cornelius and Rosewinkel [158]. Currently full-scale plants are in operation in for example Strass (Austria) and Zurich (Switzerland). The plant in Strass treats the wastewater of 200,000 population equivalents, and is equipped with a 500 m^3 sequencing batch reactor (SBR) for deammonification of reject-water originating from digested-sludge dewatering [159]. In Zurich a 1400 m^3 reactor is operational treating over 500 g N m^{-3} d^{-1} with conversions over 90% [160].

A summary of several of the lab-scale experimental studies described in literature is given in Table 5. InTable 6 results from full-scale experiences is presented.

Table 5: Summary of several lab-scale experimental studies on 1 reactor systems for autotrophic nitrogen removal described in literature

Reactor type	Volume (L)	Influent type	Inocula	HRT (d)	SRT (d)	DO (mg L⁻¹)	pH	Temperature (°C)	Removal rate (kg N m⁻³ d⁻¹)	N removal (%)	Reference
Air lift	1.8	Synthetic	Anammox + nitrifying sludge	0.42	–	0.5	7.5	30	1.5	42	[143]
RBC	44	Synthetic	RBC biomass		–	0.6		29	1.05	89	[108]
	50	Synthetic	RBC biomass	1	–	0.3	7.85	30	1.80	88	[131]
	50	Sludge liquor	RBC biomass	1	–	1.0	7.85	14	0.42	42	[131]
SBR	1.5	Reject water diluted with tapwater 1:1ᵃ	Nitrifying granular + 'Anammox'	0.5	30–110	0.5	7.5–7.9	21	0.5	78	[140]
	2.0	Synthetic	'Anammox'	1	–	<0.1	7.8	30	0.06	50	[151]
	2.5	Synthetic	OLAND	–	–	0.3–0.7	>7.4	32–34	1.1	–	[152]
	10	Sludge liquor			–	0.6		30	0.06	76	[8]
Upflow granular bed reactor	50	Sludge liquor	Nitrifying;denitrifying activated sludge		–	1.8		30	0.36	60	[156]

Moving bed biofilm reactor	4	Synthetic	Nitrifying biofilm + Anammox	-	0.5	35	0.77	89	[142]

Note: Some of the characteristics of the reactors were not reported.

[a]To reach an ammonium concentration of 0.15–0.35 g N L⁻¹.

Table 6: Overview of the operational conditions and nitrogen removal performance of several pilot and full scale 1-reactor autotrophic nitrogen removal systems

Reactor type	Volume (m³)	Influent type	pH	T (°C)	DO (mg O₂ L⁻¹)	Removal rate (kg N m⁻³ d⁻¹)	Removal efficiency (%)	Reference
SBR	500	Sludge liquor	7.05–7.10	25–30	0.3	0.6	84	[195]
	400	Sludge liquor	7.05–7.10	25–30	0.3	0.4	90	[196]
	4.1	Sludge liquor	7.4–7.6	25	0.5–1.0	0.65	90	[197]
Upflow reactor	600	Sludge liquor	8.0	30–35	2.0–3.0	1.3	75–80	[197]
MBR	Full scale	Landfill leachate	–	–	0.5	0.33	73	[198]
	Full scale	Landfill leachate	6.9	–	0.3	0.33	84	[199]

Moving bed	0.04	Sludge liquor	8.0–8.5	27	<1.0	0.5	60–70	[200]
	0.04	Sludge liquor	8–8.1	28–29	0.8–2.0	0.12–0.22	75–71	[155]
	21	Sludge liquor	7.6–8	23–27	1.2–2.6	0.38	62	[144]
	Full scale	Sludge liquor	7.8	30	3	0.35	64	[201]
	Full scale	Sludge liquor	8	27	–	0.21	72	[201]
RBC	265	Landfill leachate	8.3 (7.4–8.7)	28 (27–30)	0.7–1.0	0.15–0.26	40–70	[147]
	33	Landfill leachate	7.3	16	1.0–2.0	0.25–0.57	30–70	[86]
	240	Landfill leachate	8.1 (7.2–8.8)	14 (10–28)	0.8–1.2	1.7	30–70	[92]

Note: Some of the characteristics of the reactors were not reported.

Two strategies are possible to start-up a one-reactor autotrophic nitrogen removal system. The first method is the inoculation of nitrifying biomass into a well performing Anammox reactor and supplying air into the reactor to maintain microaerobic conditions. Otherwise, a partial nitritation reactor can be operated under oxygen limited conditions obtaining an ammonium:nitrite ratio of 1:1 before Anammox biomass is inoculated into the reactor [131] and [142]. The second strategy seems to be more appropriated since an important decrease of Anammox activity will be observed when the first method is applied [143],[151] and [161]. This high nitrifying activity can protect the Anammox bacteria from oxygen and provides them enough nitrite. The inoculation of Anammox enriched biomass in a partial nitritation reactor accelerates the start-up and allows to increase the ANR after 1 or 2 months instead of the several months or even years without inoculation [140] and [208]. Moreover, only a limited amount of Anammox biomass is necessary to start-up the CANON process with this second strategy.

Partial Nitritation and Anammox in Separate Reactors (2-Reactor Systems)

Partial Nitritation

The challenge for the first reactor in a 2-reactor system is to obtain a stable, Anammox-suited effluent, i.e. with a molar ammonium:nitrite ratio of 1:1.32 according to the stoichiometry proposed by Strous et al. [82]. In practice, however, this ratio will be closer to 1:1 in view of the desire to prevent nitrite inhibition, i.e. by providing an excess of ammonium. Up to now three types of reactors were used to achieve this: completely stirred tank reactors (CSTR), membrane bioreactors (MBR) and sequencing batch reactors (SBR). In the MBR and SBR reactor high sludge retention times are obtained (50–75 days) [162]. In the MBR the SRT is difficult to manipulate unlike in suspended growth systems which brings difficulty to suppress nitrite oxidizers even under oxygen-limited concentrations [163]. Fux et al. [164] also stated that a long-term nitrite production without nitrate accumulation can be unreliable in biofilm systems since the control of the sludge age is difficult. None of the selection criteria applied such as high free ammonia, low oxygen concentration or high ammonium loading rate led to selective suppression of nitrite oxidation in a long-term laboratory and pilot scale moving-bed biofilm reactor. For full scale applications, a CSTR or a SBR with suspended biomass is recommended.

Further, the footprint of an MBR systems is reduced due to the absence of settling tanks and the reduction in bioreactor volume due to the higher biomass concentration [163].

The possibility to obtain an Anammox-suited effluent by SHARON process was tested by van Dongen et al.[5] and Mosquera-Corral et al. [73] in a CSTR with reject water as influent at a temperature of 35 °C and a HRT and SRT of 1 day. The ammonium was for 53% oxidized to nitrite without pH control resulting in a nitrite: ammonium ratio of 1.13:1. In the subsequent Anammox reactor nitrite was therefore the limiting component. Van Hulle et al. [27] described the start-up and operation of a lab-scale SHARON reactor operated at 35 °C without pH-control. An Anammox-suited influent was obtained with synthetic influent containing an ammonium loading rate up to 1.5 kg N m^{-3} d^{-1}. Udert et al. [165] described also good SHARON performance with urine as influent. In the CSTR an ammonium:nitrite ratio of 1:1 was obtained at a HRT of 4.8 days and a pH of 9.2.

The SHARON technology is nowadays successfully used at full scale to treat effluents from sludge digesters. Full-scale SHARON reactors are currently in operation at the sludge treatment site Sluisjesdijk of the WWTP of Rotterdam and Utrecht (The Netherlands) [12]. Fux et al. [38] also operated a 2.1 m^{3}CSTR reactor in Zurich at a HRT of 1.1 days and a temperature of 30 °C without pH control. Digester effluent from two different WWTPs was tested obtaining an Anammox-suited ammonium:nitrite ratio of 1:1.32 at a pH between 6.6 and 7.2.

Although the SHARON process is successfully started up at full scale, there are still some disadvantages connected to this process.

Sludge digesters operate at high HRT values guaranteeing a stable composition of its effluents for the subsequent SHARON process (low biodegradable organic matter and a bicarbonate to ammonia molar ratio of 1). When the HRTs in the digesters are lower than usual or when industrial wastewaters are used, fluctuations of the influent composition into the SHARON reactor will occur. Therefore, operational parameters such as DO or pH must be controlled in the preceding SHARON process to obtain an optimal nitrite:ammonium ratio [26]. Another disadvantage is the limited maximum volumetric loading rate of SHARON reactor, as sludge is constantly withdrawn. To assure stable operation, the minimum HRT of a chemostat is limited to 1–1.2 days. In MBR, SBR or biofilm systems biomass is retained giving the advantage that HRT can be uncoupled from SRT and HRT lower than 1 day is possible resulting in much higher loading rates (i.e. smaller reactors with similar treatment capacity [136]). Protozoa can cause problems in the operation of a SHARON reactor mainly if real wastewater is used [5]. A possible solution is to lower the reactor pH to 6 for 2 h or to incorporate non-aerated periods. A pH-lowering can be obtained by reducing the influent flow under constant aeration. Non-aerated periods, however, clearly have a negative effect on the nitrogen conversion by nitrifiers and the SHARON reactor has to be 30% larger to maintain good nitrite formation [5]. Moreover, the required performance temperature of SHARON is higher. When the effluent of the treated stream is lower than 24 °C the maximal growth rate of AOB turns lower than that of nitrite oxidizers and ammonium is fully oxidized into nitrate [38]. Therefore, to achieve partial nitritation at temperature lower than 24 °C other strategies such as inhibition of NOB by ammonia and nitrous acid or operation at low oxygen concentrations should be applied.

Wyffels et al. [166] used a MBR as a first step of the autotrophic nitrogen removal process at low dissolved oxygen concentrations (<0.1 mg O_2 L^{-1}). The membrane had to be regularly cleaned to prevent clogging. The pH was controlled at 7.9 and the temperature was set to 35 °C, although an experiment at room temperature was conducted as well. Lowering the temperature had no significant effect on the obtained nitrite:ammonium ratio. Similarly, lowering the NH_3 concentration, and possibly lowering the NH_3 inhibition on nitrite oxidizers, had no significant effect on the obtained nitrite:ammonium ratio. This indicates that oxygen limitation is the most important operational factor. Feng et al. [167] and Xue et al. [163] also used the MBR to obtain good partial nitritation performance at low dissolved oxygen concentration. Feng et al.[167] stated that alkalinity also played an important factor to achieve a nitrite: ammonium ratio of 1.3:1 while Xue et al. [163] reported that free ammonia inhibited the nitrite oxidizers.

Ganigué et al. [168] showed that the SBR is a feasible technology to achieve stable influent for an Anammox reactor when urban landfill leachate is treated. At low pH values biological activity decreased due to an inhibitory effect by free nitrous acid and a lack of bicarbonate. On the other hand, high pH values indicated a decrease in oxygen uptake rate caused by free ammonia inhibition. As such, pH is considered to be an important factor. Udert et al. [165] used a SBR to treat urine at a temperature of 24.5 °C while varying the oxygen concentration between 2 and 4.5 mg O_2 L^{-1}. The pH at the start of the reaction cycle was 8.8 and gradually decreased to a minimum of 6 as ammonium conversion continued. At this pH ammonium conversion stopped probably because NH_3 limitation and HNO_2 inhibition obtaining an ammonium:nitrite ratio of 1. Another possible explanation is the inhibition of nitrite oxidizers by the intermediaire hydroxylamine. Yamamoto et al. [133] applied the partial nitritation and Anammox process to treat swine wastewater digester liquor. They observed that a stable conversion of ammonia into nitrite of 58% could be reached in a biofilm reactor due to inhibition of free ammonia and free nitrous acid. The inhibition of free ammonia was also brought forworth by Liang et al. [169] and Qiao et al. [170] who treated landfill leachate and digested liquid manure, respectively in a biofilm reactor achieving a nitrite: ammonia molar ratio near 1.3.

Table 7: Summary of several experimental studies concerning partial nitritation in view of coupling with an Anammox reactor described in literature

Process	Reactor type	Volume (L)	Influent type	pH	Temperature (°C)	DO (mg L⁻¹)	SRT (d)	HRT (d)	N load (kg N m⁻³ d⁻¹)	NO₂⁻:NH₄⁺ ratio	Nitrate in the effluent (%)	Reference
SHA-RON	CSTR	2100	Sludge digester effluent from WWTP	6.6–7.2	29	2.7	1.05–1.18	1.05–1.181	0.56	1.4	Negligible	[38]
SHA-RON	CSTR	2	Synthetic	7.1	35	–	1–1.5	1.54	1.5	1	–	[27]
SHA-RON	CSTR	10	Sludge digester effluent from WWTP	6.7	35	–	1	1	1.2	0.74	113	[5]
SHA-RON	CSTR	2.8	Urine	9.2	30	2.5–4	4.8	4.8	1.580	1	Negligible	[165]
SHA-RON	CSTR	3.2	Digested effluent of fish canning	7.5	35	>2	1	1	0.1	1	No nitrate	[73]

Partial nitritation	SBR	7.5	Urine	6–8.8	24.5	2–4.5	>30	4	0.560	1	Negligible	[165]
Partial nitritation	SBR	20	Landfill leachate	6.8–7.1	36	2	3–7	1.5	1.5	0.6–1.5	<5	[168]
Partial nitritation	Biofilm	10.8 + 2.5	Pre-filtered digested liquor of swine water[a]	–	25	5	13	1	1.0	1.38	<5[b]	[133]
Partial nitritation	Upflow fixed bed biofilm reactor	11	Landfill leachate	8.4	30	0.8–2.3	–	Varying	0.27–1.2	58.3		[169]
Partial nitritation	Column with PEG carrier	8.0	Digested liquid manure	7.5–8	30	2.5–6.5	–	–	3.8	1.22	2.3	[170]
Partial nitritation	MBR	1	Synthetic	8	35	<0.6	35	0.23	–	1.30	Trace	[167]

| Partial nitritation | MBR | 14 | Synthetic | 8 | 35 | 0.3–0.5 | – | 0.67 | 0.450 | 1 | – | [163] |
| Partial nitritation | MBR | 1.5 | Pre-filtered Sludge digester effluent of WWTP | 7.9 | 30 | <0.2 | Varying | 0.58–1 | 0.73–1.45 | 1.13 | – | [166] |

Note: Some characteristics of the reactors were not reported.

aDiluted with tap water to achieve a NLR of 1 kg N m^{-3} d^{-1}.

bConversion efficiency of ammonia in nitrate.

In the different experiments described above, different conditions were used to favour the growth of ammonium oxidizers over nitrite oxidizers in order to produce an Anammox-suited-influent. Four principles can be distinguished: the operation of the reactor at low dissolved oxygen concentration (<0.5 mg O_2 L^{-1}), the operation of the reactor at high pH (7.5–8.5), which increases the ammonia availability and decreases the nitrous acid availability, the operation of the reactor at high temperature (>25 °C), a limited nitrification time which stops ammonium oxidation before its depletion and the presence of a bicarbonate limitation which stops nitrification. Table 7 summarizes several experiments described in literature.

Anammox

The practical application of the Anammox process is still limited by its long start-up periods (up to 1 year) due to the very low growth rate and low cell yield of Anammox organisms. Loss of a fraction of the sludge due to wash-out with the effluent could further augment the start-up period. Hence, it is essential to use a reactor with high biomass retention. The cultivation of slow-growing microorganisms relies mostly on the ability of biomass to form biofilms or aggregates such as flocs or granules [100]. So far a large range of bioreactor types have been evaluated for the enrichment of Anammox bacteria: fixed bed reactors, fluidized-bed-reactors, UASB-reactors, SBR, gas-lift reactors [136] and [171]. Among them, the SBR was accepted for Anammox enrichment for its simplicity, efficient biomass retention, homogeneity of mixture in the reactor, stability and reliability for a long period of operation, stability under substrate-limiting conditions and high nitrogen conversions [5], [43] and [82]. The SBR reached a biomass retention of 90% which was 1.4 times more than in a fluidized bed reactor [172]. Strous et al. [172] started up the Anammox process in a fixed-bed and fluidized bed reactor with glass and sand particles as carriers but could not prevent biomass loss due to floating sludge caused by entrapped gas bubbles. The same situation occurred in the gas lift reactor at increased nitrogen removal rate [173]. Dapena-Mora et al. [173] stated that mechanical stirring in a SBR could be more effective to eliminate the gas entrapped in the granules compared to the less abrasive stress in a gas-lift. Further, also the application of non-woven fibers can incease the biomass retention as several experiments with nonwoven fibers demonstrated a short start-up time and high nitrogen removal rates [98] and [101].

An alternative for obtaining full biomass retention in Anammox systems might be the use of membrane bioreactors (MBR). Unlike the reactors with granular biomass, the MBR enables cultivation of slow growing bacteria with biomass retention and without a selection on settling ability. van der Star [100] pointed out that the MBR reactor is a more powerful tool for Anammox research as high production of almost pure suspended anammox cells could be obtained avoiding the diffusion limitations within flocs or granules. A membrane SBR which is a combination of a SBR and a biofilm system was applied by Trigo et al. [134]achieving a high nitrogen removal rate. Wang et al. [176] used a stirrer in the MBR to make the Anammox bacteria suspended as free cells and a more homogeneous distribution of substrates and biomass can be achieved. However, for full-scale applications biofilm- or granular-based bioreactors are preferable over MBRs since anammox bacteria easily form sludge granules or biofilms obtaining a high biomass concentration in the reactor on a simple and economical way. Further, fouling of the membrane system could occur. The operation costs due to backwashing (high energy consumption) or external cleaning with chemicals are inevitable in engineering practice [134] and [177]. Moreover, wastewater always contains a certain amount of solids which are also retained in a MBR reactor. This accumulation of solids could decrease the activity in a full-scale MBR-based anammox process [133].

A summary of the experimental studies described in literature is given in Table 8. From these studies the potential of the Anammox process can be seen since total nitrogen removal rates up to 26 kg N m^{-3} d^{-1} in a fixed bed reactor fed with synthetic wastewater [174]. In contrast, the nitrogen removal rate is not so high in engineering. A possible explanation of the lower nitrogen removal rates in pilot plants is the limited availability of substrate in real waste waters. The efficiency of biomass retention is another factor which determines the maximum conversion while in biofilm reactors, nitrite fux to the biofilm is another potential limitation. Isaka et al. [128] stated that HRT has an influence on the nitrogen removal rates. Under appropriate nitrite and ammonium concentrations nitrogen conversion rates can be increased by decreasing the HRT. Wyffels et al. [136] stated that the maximum nitrogen removal rate of Anammox organisms is limited by the growth rate of ammonium oxidizers in the SHARON process since a minimum HRT of 1 day is needed.

The concentration of nitrite during the start-up is of crucial importance for growth: a too low amount will result in substrate limitation and thus slower growth, while concentration above 20 mg N L^{-1} can already lead to inhibition. As such, nitrite levels could increase even more leading to complete process failure. Start-up of Anammox reactors is often characterized by a gradual increase of nitrite concentration in the influent. The nitrite:ammonium ratio in the influent reaches 1 although often an excess of ammonium is used allowing a lower overall nitrogen removal efficiency but guaranteeing a more stable process. Since the Anammox process is anaerobic, the absence of oxygen is an essential step especially during the start-up of reactors [173]. Further, the impact of variability in real streams on the performance of Anammox in full-scale reactors is not well understood [178] yet.

To fasten up the start-up period, Anammox biomass is often used as inocula of Anammox reactors. The fast start-up time of 14 days in a SBR reactor by Sliekers et al. [143] was due to the inoculation of the reactor with fully active Anammox sludge. For the other reactors start-up time was significantly higher. Sequential addition of the pre-enriched Anammox sludge was also selected as a strategy for the engineering practice in the Netherlands [175]. The 10 L lab scale reactor was directly scaled up to a full scale reactor of 70 m^3 reactor. This reactor was initiated in Rotterdam in 2002 and the start-up took nearly 3.5 years. Now stable operation reached a nitrogen removal rate of 9.5 kg N m^{-3} d^{-1}[175] (Table 8).

Table 8: Overview of operational parameters and nitrogen removal of a selection of studies on Anammox

Reactor type	Volume (L)	Influent type	Inoccula	HRT (d)	pH	T (°C)	NLR (g N L⁻¹ d⁻¹)	NRR (g N L⁻¹ d⁻¹)	SNR (g N g VSS⁻¹ d⁻¹)	N removal (%)	Reaction N ratio (NH₄⁺:NO₂⁻:NO₃⁻)	Reference
SBR	20	Anaerobic digester centrate	Anammox	–	7.5–7.8	35	0.380	0.2ª	–	85	1:1.20:0.13	[202]
SBR	1600	Digester effluent from WWTP	Anammox biomass	–	7.45–7.59	30.4–31.8	0.650ª (2.6)	0.56–0.64 (2.4)	0.075	85–99	1:1.38:0.32	[38]
SBR	1	Anaerobic digester supernatant	Anammox biomass	1	7.5	19–21	0.28	0.08	0.13	69	–	[206]
SBR	1.2	Synthetic	Digested sludge of WWTP	0.1	7.5–8	30	2.7	2.0	–	–	1:1.08:-	[203]
SBR	1	Synthetic	Anammox granular sludge	0.625	7.8–8	35	1.0	0.7	0.65	78	1:1.11:0.2	[173] and [204]
SBR	1	Synthetic	Activated sludge	1	7.8	35–36	0.6	0.6	0.21	–	–	[205]
SBR	10	Digester liquor	Anammox	1	7–8	30–37	1.0	0.55–0.95	0.18	–	1:1.03:-	[5]

Reactor	Volume	Influent	Biomass	1.8	7.5–8.2	35	0.34–0.67	0.3	0.44	40–80		Ref
SBR	3	Effluent from a fish-canning industry pre-digested	Anammox sludge	–	7.5–8.2	35	0.34–0.67	0.3	0.44	40–80	–	[140]
Air lift	70,000	Digester effluent from WWTP	Activated sludge + Anammox biomass	–	7–8	30–40	9.5	9	–	–	1:1.32:0.25	[175]
Airlift	7 + 3	Synthetic	Anammox granular sludge	1	7.9–8.1	30	2.3	2.0	1.15	88	1:1.28:0.26	[173]
Airlift	1.8	Synthetic		0.28	7.5	30	10.7	8.9	–	–	–	[143]
MBR	1.5	Digester effluent from WWTP	Anammox biomass	0.75–1.1		20–30	0.65–1.1	0.55	–	82	1:1.05:0,20	[136]
MBR	5	Synthetic	Anammox granular sludge	1	8	35	0.74	0.71	0.45	73.6	1:1.22:0.22	[134]
MBR	4.8	Synthetic	Nitrify-ing and denitrifying Activated sludge	2	8	35	–	–	0.35	90	1:1.15:–	[176]
CSTR	0.73	Liquid manure digestor liquor	Anammox granular sludge	0.2	7.5	30	3.73	2.60	–	77	1:1.20:0.22	[170]

Fixed bed	0.8	Synthetic	Activated WTP-sludge	0.06–0.3	7.0–7.5	37	0.1–9.4	6.2	–	–	–	
	0.8	Synthetic	Concentrated activated WWTP-sludge	0.01–0.3	7.0–7.5	37	0.1–58.5	26.0	1.6	–	1:1.2:0.33	[174]
UF	36	Landfill leachate	Activated sludge + Anammox sludge		7.5–8	30		0.11	–	62	1:1.09:0.07	[169]
UF	2.85	Liquid manure	Anammox sludge	1	7.2–7.6	35	0.39	0.22		55	1:1.67:0.53	[133]
UF	2.85	Synthetic	Anammox sludge	1	7.2–7.6	35	0.67	0.5		73	1:1.26:0.33	[133]
FB	2.5	Synthetic	Denitrifying sludge	0.9–1.75	8	36	2.0	1.8	0.18	–	1:1.5:–	[172]
FB	2.5	Effluent of digested WWT sludge	Denitrifying sludge	0.14–11	8	36	2.5	1.5	0.15	–	1:0.55:–	[172]
Continuous flow	19.4	Digested effluent from WWTP	Anammox biomass	–	7.5–7.8	35	0.14–0.38	0.33	–	85–91	1:1.21:0.13	[178]
ABF	0.2	Synthetic	Anammox biomass	0.03	7.2	20–22	12	8.1	–	–	–	[128]
ABF	0.2	Synthetic	Anammox biomass	0.06	7.2	37	19.1	11.5	–	–	–	[128]

UBF	200	Effluent of partial nitritation		0.38	7.5	30	7.0	6.4					[207]
UBF	1.2	Synthetic	Digested sludge of WWTP	0.12	7.5–8	30	2.5	2.0	–	–		1:1.37:–	[203]
UASB	1 + 0.5	Piggery waste	Granular sludge	5	8.2–8.5	35	1.02[b]	0.66	0.08	80		1:1.65:0	[70]
UASB	1 + 0.5	Piggery waste	Granular sludge	5	8.2–8.5	35	0.84[b]	0.59	0.06	82		1:1.13:0	[70]

[a]A nitrite surplus in the effluent of the nitritation reactor is balanced by adding raw digester effluent.

[b]With addition of synthetic nitrite.

THE FUTURE OF AUTOTROPHIC NITROGEN REMOVAL: TOWARDS FULL-SCALE APPLICATIONS WITH LOW START-UP TIMES

Autotrophic nitrogen removal offers a useful and sustainable alternative for treating highly loaded nitrogen streams with an unfavourable carbon to nitrogen (C N^{-1}) ratio. The process has already been studied extensively on lab-scale and pilot-scale by research groups around the world. The resulting engineering aspects and practical implementation of these studies were reviewed in this contribution. An Anammox-suited effluent can be produced by selection of the substrate availability and the appropriate temperature, pH and oxygen conditions in the partial nitritation reactor. Careful control of this reactor as well as the limitation of inhibiting factors in the Anammox reactor are essential for the successful operation of the combined process.

These efforts on lab-scale resulted in a growing number of full-scale applications. For the further development of autotrophic nitrogen removing processes, research should be conducted towards fast start-up strategies and sustainable control of the process. Further research into the fundamental Anammox behaviour will certainly help to improve this operational control.

ACKNOWLEDGMENTS

Peter Vanrolleghem is Canada Research Chair in Water Quality Modelling.

The authors would like to thank Ana Dapena-Mora, Mike Jetten, Ramon Mendez, Gurkan Sin, Sammy Van Den Broeck, Brecht Donckels, Jo Maertens, Kris Villez, Eveline Volcke, Mark van Loosdrecht, Willy Verstraete, and Stijn Wyffels for the numerous discussions on the subject of (autotrophic) nitrogen removal.

The authors would like to thank the European Union and the Institute for the Promotion of Innovation by Science and Technology in Flanders (IWT) for the financial support (EU project number EVK1-CT2000-00054; IWT project number 80126).

REFERENCES

1.   Metcalf & Eddy, Inc., Revised by G. Tchobananoglous, F.L. Burton, Wastewater engineering: treatment, disposal and reuse. McGraw-Hill, McGraw-Hill series in water resources and environmental engineering, New York, USA, 2003.

2.   U. Wiesmann, Biological nitrogen removal from wastewater, Advances in Biochemical Engineering/Biotechnology 51 (1994) 113–154.

3.   A. Mulder, A.A. van de Graaf, L.A. Robertson, J.G. Kuenen, Anaerobic ammonium oxidation discovered in a denitrifying fluidized bed reactor, FEMS Microbiology Ecology 16 (1995) 177–184.

4.   E. Broda, Two kinds of lithotrophsmissing in nature, Zeitschrift fur Allgemeine Mikrobiologie 17 (1977) 491–493.

5.   U. van Dongen, M.S.M. Jetten, M.C.M. van Loosdrecht, The SHARON®- Anammox® process for treatment of ammonium rich wastewater, Water Science & Technology 44 (1) (2001) 153–160.

6.   M.S.M. Jetten, S.J. Horn, M.C.M. van Loosdrecht, Towards a more sustainable wastewater treatment system, Water Science & Technology 35 (9) (1997) 171–180.

7.   M.J. Kampschreur, H. Temmink, R. Kleerebezem, M. Jetten, M.C.M. van Loosdrecht, Nitrous oxide emission during wastewater treatment,Water Research 43 (2009) 4093–4103.

8.   Y.-H. Ahn, Sustainable nitrogen elimination biotechnologies: a review, Process Biochemistry 41 (2006) 1709–1721.

9.   E. Plaza, J. Trela, L. Gut, M. Löwén, B. Szatkowska, Deammonification process for treatment of ammonium

rich wastewater. In: Integration and optimisation of urban sanitation systems, Joint Polish-Swedish Reports, No. 10. TRITA-LWR REPORT 3004, pp. 77–87, Gdansk, Poland, March 23–25, 2003.

10.   S.E. Vlaeminck, Biofilm and granule applications for one-stage autotrophic nitrogen removal, Phd Thesis, Ghent University, Belgium, 2009.

11.   A. Mulder, The quest for sustainable nitrogen removal technologies, Water Science & Technology 48 (1) (2003) 67–75.

12.   J.W. Mulder, M.C.M. van Loosdrecht, C. Hellinga, R. van Kempen, Full-scale application of the Sharon process for the treatment of rejection water of digested sludge dewatering, Water Science & Technology 43 (11) (2001) 27–134.

13.   M.M. Clabaugh, Nitrification of landfill leachate by biofilm columns, MSc Thesis, Virginia Polytechnic Institute and State University, USA, 2001, 43 p.

14.   P. Ilies, D. Mavinic, The effect of decreased ambient temperature on the biological nitrification and denitrification of a high ammonia landfill leachate, Water Research 35 (2001) 2065–2072.

15.   T. Feyaerts, D. Huybrechts, R. Dijkmans, 2002, Best beschikbare technieken voor mestverwerking, Technical report, Vlaams Instituut voor Technologisch Onderzoek VITO (in Dutch).

16.   J. Carrera, J. Baeza, T. Vicent, J. Lafuente, Biological nitrogen removal of highstrength ammonium industrial wastewater with two-sludge system, Water Research 37 (2003) 4211–4221.

17.   S. Murat, G. Insel, N. Artan, D. Orhon, Peformance evaluation of SBR treatment for nitrogen removal from tannery wastewater, in: IWA Speciality Symposium on Strong Nitrogenous and Agro-Wastewater, vol. 2, Seoul, Korea, June 11–13, 2003, pp. 598–605.

18. J. Keller, K. Subramaniam, J. Gösswein, P. Greenfield, Nutrient removal from industrial wastewater using single tank sequencing batch reactors, Water Science & Technology 35 (6) (1997) 137–144.

19. U. Abeling, C.F. Seyfried, Anaerobic-aerobic treatment of potato-starch wastewater, Water Science & Technology 28 (2) (1993) 165–176.

20. J. Campos, A. Sanchez, A. Mosquera-Corral, R. Mendez, J. Lema, Coupled BAS and anoxic USB system to remove urea and formaldehyde from wastewater, Water Research 37 (2003) 3445–3451.

21. J. Garrido, R. Mendez, J. Lema, Simultaneous urea hydrolysis, formaldehyde removal and denitrification in a multifed upflow filter under anoxic and anaerobic conditions, Water Research 35 (2001) 691–698.

22. A. Carucci, A. Chiavola, M. Majone, E. Rolle, Treatment of tannery wastewater in a sequencing batch reactor, Water Science & Technology 40 (1) (1999) 253–259.

23. M. Ros, A. Gantar, Possibilities of reduction of recipient loading of tannery wastewater in Slovenia, Water Science & Technology 37 (8) (1998) 145–152.

24. A. Abeliovich, Transformations of ammonia and the environmental impact of nitrifying bacteria, Biodegradation 3 (1992) 255–264.

25. Y. Peng, G. Zhu, Biological nitrogen removal with nitrification and denitrification via nitrite pathway, Applied Microbiology and Biotechnology 73 (2006) 15–26.

26. E.I.P. Volcke, M.C.M. van Loosdrecht, P.A. Vanrolleghem, Controlling the nitrite:ammonium ratio in a SHARON reactor in view of its coupling with an Anammox process, Water Science & Technology 53 (4–5) (2006) 45–54.

27. S.W.H. Van Hulle, S. Van Den Broeck, J. Maertens, K. Villez, B.M.R. Donckels, G. Schelstraete, E.I.P. Volcke, P.A. Vanrolleghem, Construction, start-up and operation of a continuously aerated laboratory-scale SHARON reactor in

view of coupling with an Anammox reactor, Water SA 31 (2005) 327–334.

28. I. Suzuki, U. Dular, S.C. Kwok, Ammonia or ammonium ion as substrate for oxidation by Nitrosomonas europaea cells and extracts, Journal of Bacteriology 120 (1974) 556–558.

29. A.C. Anthonisen, R.C. Loehr, T.B.S. Prakasam, E.G. Srinath, Inhibition of nitri- fication by ammonia and nitrous acid, Journal of Water Pollution Control Federation 48 (1976) 835–852.

30. S.W.H. Van Hulle, E.I.P. Volcke, J. López Teruel, B. Donckels, M.C.M. van Loosdrecht, P.A. Vanrolleghem, Influence of temperature and pH on the kinetics of the SHARON nitritation process, Journal of Chemical Technology & Biotechnology 82 (2007) 471–480.

31. Turk, D.S. Mavinic, Maintaining nitrite buildup in a system acclimated to free ammonia, Water Research 23 (1989) 1383–1388.

32. T.B.S. Prakasam, R.C. Loehr, Microbial nitrification and denitrification in concentrated wastes, Water Research 6 (1972) 859–869.

33. S. Hawkins, K. Robinson, A. Layton, G. Sayler, Limited impact of free ammonia on Nitrobacter spp. Inhibition assessed by chemical and molecular techniques, Bioresource Technology 101 (2010) 4513–4519.

34. C. Hellinga, A. Schellen, J.W. Mulder, M.C.M. van Loosdrecht, J.J. Heijnen, The SHARON process: an innovative method for nitrogen removal from ammonium-rich wastewater, Water Science & Technology 37 (1998) 135–142.

35. B. Wett, W. Rauch, The role of inorganic carbon limitation in biological removal of extremely ammonia concentrated wastewater, Water Research 37 (2002) 1100–1110.

36. A. Guisasola, S. Petzet, J.A. Baeza, J. Carrera, J. Lafuente, Inorganic carbon limitations on nitrification: Experimental assessment and modelling, Water Research 41 (2007) 277–286.

37. S. Tarre, M. Beliavski, N. Denekamp, A. Gieseke, D. de Beer, M. Green, High nitrification rate at low pH in a fluidized bed reactor with chalk as the biofilm carrier, Water Science & Technology 49 (11–12) (2004) 99–105.

38. C. Fux, M. Boehler, P. Huber, I. Brunner, H. Siegrist, Biological treatment of ammonium-rich wastewater by partial nitrification and subsequent anaerobic ammonium oxidation (anammox) in a pilot plant, Journal of Biotechnology 99 (2002) 295–306.

39. R. Ganigué, J. Gabarro, A. Sànchez-Melsió, M. Ruscalleda, H. López, X. Vila, J. Colprim, M.D. Balaguer, Long-term operation of a partial nitritation pilot plant treating leachate with extremely high ammonium concentration prior to an anammox process, Bioresource Technology 100 (2009) 5624–5632.

40. O. Nowak, K. Svardal, H. Kroiss, The impact of phosphorus deficiency on nitrification—case study of a biological pretreatment plant for rendering plant effluent, Water Science & Technology 34 (1–2) (1996) 229–236.

41. C. Grunditz, G. Dalhammar, Development of nitrification inhibition assays using pure cultures of Nitrosomonas and Nitrobacter, Water Research 35 (2001) 433–440.

42. C. Hellinga, M.C.M. van Loosdrecht, J.J. Heijnen, Model based design of a novel process for nitrogen removal from concentrated flows, Mathematical and Computer Modelling of Dynamical Systems 5 (1999) 351–371.

43. M.S.M. Jetten, M. Strous, K.T. van de Pas-Schoonen, J. Schalk, U.G.J.M. van Dongen, A.A. Van De Graaf, S. Logemann, G. Muyzer, M.C.M. van Loosdrecht, J.G. Kuenen, The anaerobic oxidation of ammonium, FEMS Microbiology Reviews 22 (1999) 421–437.

44. W. Helder, R.T.P. De Vries, Estuarine nitrite maxima and nitrifying bacteria (Ems-Dollard estuary), Netherlands Journal of Sea Research 17 (1983) 1–18.

45. G. Knowles, A.L. Downing, M.J. Barrett, Determination of kinetic constants for nitrifying bacteria in mixed culture,

with the aid of electronic computer, Journal of General Microbiology 38 (1965) 263–278.

46. F.E. Stratton, P.L. Mc Carty, Microbiological aspects of ammonia oxidation of swine waste, Canadian Journal of Microbiology 37 (1967) 918–923.

47. T. Yamamoto, K. Takaki, T. Koyama, K. Furukawa, Novel partial nitritation treatment for anaerobic digestion liquor of swine wastewater using swimbed technology, Journal of Bioscience and Bioengineering 102 (6) (2006) 497–503.

48. S. Philips, H.J. Laanbroek,W. Verstraete, Origin, causes and effects of increased nitrite concentrations in aquatic environments, Re/Views in Environmental Science and Bio/ Technology 1 (2002) 115–141.

49. K. Hanaki, C. Wantawin, S. Ohgaki, Nitrification at low-levels of dissolvedoxygen with and without organic loading in a suspended-growth reactor, Water Research 24 (1990) 297–302.

50. J.H. Hunik, J. Tramper, R.H. Wijffels, A strategy to scale-up nitrification processes with immobilized cells of Nitrosomonas europaea and Nitrobacter agilis, Bioprocess Engineering 11 (1994) 73–82.

51. D. Barnes, P.J. Bliss, Biological Control of Nitrogen in Wastewater Treatment, E.&F.N. Spon, London, UK, 1983, p. 365.

52. G. Ciudad, O. Rubilar, P. Munoz, G. Ruiz, R. Chamy, C. Vergara, D. Jeison, Partial nitrification of high ammonia concentration wastewater as a part of a shortcut biological nitrogen removal process, Process Biochemistry 40 (2005) 1715–1719.

53. E.V. Münch, P. Lant, J. Keller, Simultaneous nitrification and denitrification in bench-scale sequencing batch reactors, Water Research 30 (1996) 277–284.

54. R. Manser, W. Gujer, H. Siegrist, Consequences of mass transfer effects on the kinetics of nitrifiers, Water Research 39 (2005) 4633–4642.

55.   Y. Peng, Y. Chen, S. Wang, C. Peng, M. Liu, X. Song, Y. Cui, Nitrite accumulation by aeration controlled in sequencing batch reactors treating domestic wastewater, Water Science & Technology 50 (10) (2004) 35–43.

56.   I. Jubany, J. Lafuente, J.A. Baeza, J. Carrer, Total and stable washout of nitrite oxidizing bacteria from a nitrifying continuous activated sludge system using automatic control based on Oxygen Uptake Rate measurements, Water Research 43 (2009) 2761–2772.

57.   T. Hidaka, H. Yamada, M. Kawamura, H. Tsuno, Effect of dissolved oxygen conditions on nitrogen removal in continuously fed intermittent-aeration process with two tanks, Water Science & Technology 45 (2002) 181–188.

58.   Y. Hyungseok, A. Kyu-Hong, L. Kwang-Hwan, K. Youn-Ung, S. Kyung-Guen, Nitrogen removal from synthetic wastewater by simultaneous nitrification and denitrification (SND) via nitrite in a intermittently-aerated reactor, Water Research 33 (1999) 145–154.

59.   A. Katsogiannis, M. Kornaros, G. Lyberatos, Enhanced nitrogen removal in SBRs by bypassing nitrate generation accomplished by multiple aerobic/anoxic phase pairs, Water Science & Technology 47 (11) (2003) 53–59.

60.   R. Blackburne, Z. Yuan, J. Keller, Demonstration of nitrogen removal via nitrite in a sequencing batch reactor treating domestic wastewater, Water Research 42 (2008) 2166–2176.

61.   G. Sin, K. Villez, P.A. Vanrolleghem, Application of a model-based optimization methodology for nutrient removing SBRs leads to falsification of the model, Water Science & Technology 53 (4–5) (2006) 95–103.

62.   L. Yang, J.E. Alleman, Investigation of batchwise nitrite build-up by an enriched nitrification culture, Water Science & Technology 26 (5–6) (1992) 997–1005.

63.   S.S. Hu, Acute substrate-intermediate-product related inhibition of nitrifiers, MSc Thesis, School of Civil Engineering, Purdue University, West Lafayette, Indiana, USA, 1990.

64. D. Castignetti, H.B. Gunner, Differential tolerance of hydroxylamine by an Alcaligenes sp., a heterotrophic nitrifier, and by Nitrobacter agilis, Canadian Journal of Microbiology 28 (1982) 148–150.

65. R. Stüven, M. Vollmer, E. Bock, The impact of organic matter on nitric oxide formation by Nitrosomonas europaea, Archives of Microbiology 158 (1992) 439–443.

66. E. Bock, H.P. Koops, H. Harms, Cell biology of nitrifying bacteria, in: J.I. Prosser (Ed.), Nitrification, IRL, Oxford, 1986, pp. 17–38.

67. R. Van Kempen, J.W. Mulder, M.C.M. van Loosdrecht, Overview: full scale experience of the SHARON process for treatment of rejection water of digested sludge dewatering, Water Science & Technology 44 (2001) 145–152.

68. A. Pollice, V. Tandoi, C. Lestingi, Influence of aeration and sludge retention time on ammonium oxidation to nitrite and nitrate, Water Research 36 (10) (2002) 2541–2546.

69. A. Pollice, T. Valter, L. Carmela, Influence of aeration and sludge retention time on ammonium oxidation to nitrite and nitrate, Water Research 36 (10) (2002) 2541–2546.

70. Y.-H. Ahn, I.-S. Hwang, K.-S. Min, ANAMMOX and partial denitritation in anaerobic nitrogen removal from piggery waste, Water Science & Technology 49 (5–6) (2004) 145–153.

71. B. Molinuevo, M.C. Garcia, D. Karakashev, I. Angelidaki, Anammox for ammonia removal from pig manure effluents: effect of organic matter content on process performance, Bioresource Technology 100 (2009) 2171–2175.

72. Q.X. Yang, M. Yang, S.J. Zhang, W.Z. Lv, Treatment of wastewater from a monosodium glutamate manufacturing plant using successive yeast and activated sludge systems, Process Biochemistry 40 (2005) 2483–2488.

73. A. Mosquera-Corral, F. González, J.L. Campos, R. Méndez, Partial nitrification in a SHARON reactor in the presence of salts and organic carbon compounds, Process Biochemistry 40 (2005) 3109–3118.

74. A. Zepeda, A.-C. Texiera, E. Razo-Flores, J. Gomeza, Kinetic and metabolic study of benzene, toluene and m-xylene in nitrifying batch cultures,Water Research 40 (2006) 1643–1649.

75. G. Camilla, G. Lena, D. Gunnel, Comparison of inhibition assays using nitrogen removing bacteria: application to industrial wastewater, Water Research 32 (10) (1998) 2995–3000.

76. A.M. Eilersen, M. Henze, L. Kloft, Effect of volatile fatty acids and trimethylamine on nitrification in activated sludge, Water Research 28 (1994) 1329–1336.

77. T.G. Tomlinson, A.G. Boon, C.N.A. Trotman, Inhibition of nitrification in the activated sludge process of sewage disposal, Journal of Applied Bacteriology 29 (1966) 266–291.

78. J. Surmacz-Gorska, K. Gernaey, C. Demuynck, P.A. Vanrolleghem, W. Verstraete, Nitrification monitoring in activated sludge by oxygen uptake rate (OUR) measurements, Water Research 30 (1996) 1228–1236.

79. Y. Peng, X. Song, C. Peng, J. Li, Y. Chen, Biological nitrogen removal in SBRs bypassing nitrate generation accomplished by chlorination and aeration time control, Water Science & Technology 49 (5–6) (2004) 295–300.

80. A.A. van de Graaf, A. Mulder, P. De Bruijn, M.S.M. Jetten, L.A. Robertson, J.G. Kuenen, Anaerobic oxidation of ammonia is a biologically mediated process, Applied & Environmental Microbiology 61 (1995) 1246–1251.

81. A.A. van de Graaf, P. De Bruijn, L.A. Robertson, M.S.M. Jetten, J.G. Kuenen, Autotrophic growth of anaerobic ammonium oxidation on the basis of 15N studies in a fluidized bed reactor, Microbiology 143 (1997) 2415–2421.

82. M. Strous, J.J. Heijnen, J.G. Kuenen,M.S.M. Jetten, The sequencing batch reactor as a powerful tool for the study of slowly growing anaerobic ammoniumoxidizing microorganisms, Applied Microbiology & Biotechnology 50 (1998) 589–596.

83. A.A. van de Graaf, P. De Bruijn, L.A. Robertson, M.S.M. Jetten, J.G. Kuenen, Autotrophic growth of anaerobic ammonium-oxidizing microorganisms in a fluidized bed reactor, Microbiology 142 (1996) 2187–2196.

84. M.S.M. Jetten, M. Wagner, J. Fuerst, M. van Loosdrecht, G. Kuenen, M. Strous, Microbiology and application of the anaerobic ammonium oxidation ('ANAMMOX') process, Current Opinion in Biotechnology 12 (2001) 283–288.

85. M. Schmid, B. Maas, A. Dapena, K. van de Pas-Schoonen, J. van de Vossenberg, B. Kartal, L. Van Niftrik, I. Schmidt, I. Cirpus, J.G. Kuenen, M. Wagner, J.S. Sinninghe Damsté, M.M.M. Kuypers, N.P. Revsbech, R. Mendez, M.S.M. Jetten, M. Strous, Biomarkers for in situ detection of anaerobic ammonium oxidizing (Anammox) bacteria, Applied & Environmental Microbiology 71 (2005) 1677–1684.

86. H. Siegrist, S. Reithaar, G. Koch, P. Lais, Nitrogen loss in a nitrifying rotating contactor treating ammonium-rich wastewater without organic carbon, Water Science & Technology 38 (8–9) (1998) 241–248.

87. R. Chouari, D. Le Paslier, P. Daegelen, P. Ginestet, J. Weissenbach, A. Sghir, Molecular evidence for novel planctomycete diversity in a municipal wastewater treatment plant, Applied & Environmental Microbiology 69 (2003) 7354–7363.

88. M. Kuypers, A.O. Sliekers, G. Lavik, M. Schmid, B.B. Jorgensen, J.G. Kuenen, J.S. Sinninghe Damsté, M. Strous, M.S.M. Jetten, Anaerobic ammonium oxidation by anammox bacteria in the Black Sea, Nature 422 (2003) 608–611.

89. M. Trimmer, J.C. Nicholls, B. Deflandre, Anaerobic ammonium oxidation measured in sediments along the Thames Estuary, United Kingdom, Applied & Environmental Microbiology 69 (2003) 6447–6454.

90. T. Dalsgaard, B. Thamdrup, Factors controlling anaerobic ammonium oxidation with nitrite in marine sediments,

Applied & Environmental Microbiology 68 (2002) 3802–3808.

91. M. Strous, J.A. Fuerst, E.H.M. Kramer, S. Logemann, G. Muyze, K.T. Van De PasSchoonen, R. Webb, J.G. Kuenen, M.S.M. Jetten, Missing litotroph identified as new plantomycete, Nature 400 (1999) 446–449.

92. M. Schmid, K. Walsh, R. Webb, W.I.C. Rijpstra, K. van de Pas-Schoonen, M.J. Verbruggen, T. Hill, B. Moffett, J. Fuerst, S. Schouten, J.S. Sinninghe Damsté, J. Harris, P. Shaw, M. Jetten, M. Strous, Candidatus "Scalindua brodae", sp. nov., Candidatus "Scalindua wagneri", sp. nov., two new species of anaerobic ammonium oxidizing bacteria, Systematic & Applied Microbiology 26 (2003) 529–538.

93. K. Furukawa, P.K. Lieu, H. Tokitoh, T. Fujii, Development of single-stage nitrogen removal using anammox and partial nitritation (SNAP) and its treatment performances, Water Science & Technology 53 (6) (2006) 83–90.

94. J. Schalk, H. Oustad, J.G. Kuenen, M.S.M. Jetten, The anaerobic oxidation of hydrazine: a novel reaction in microbial nitrogen metabolism, FEMS Microbiology Letters 158 (1998) 61–67.

95. J.G. Kuenen, M.S.M. Jetten, Extraordinary anaeribic ammonia-oxidizing bacteria, ASM News 67 (2001) 456–463.

96. S. Schouten, M. Strous, M.M.M. Kuypers, W.I.C. Rijpstra, M. Baas, C.J. Schubert, M.S.M. Jetten, J.S. Sinninghe Damsté, Stable carbon isotopic fractionations associated with inorganic carbon fixation by anaerobic ammonium-oxidizing bacteria, Applied & Environmental Microbiology 70 (2004) 3785–3788.

97. M.R. Lindsay, R. Webb, M. Strous, M.S.M. Jetten, M.K. Butler, R.J. Forde, J.A. Fuerst, Cell compartimentalisation in planctomycetes: novel types of structural organisation for the bacterial cell, Archives of Microbiology 175 (2001) 413–429.

98. L.A. Van Niftrik, J.A. Fuerst, J.S. Sinninghe Damsté, J.G. Kuenen, M.S.M. Jetten, M. Strous, The anammoxosome: an

intracytoplasmic compartment in anammox bacteria, FEMS Microbiology Letters 233 (2004) 7–13.

99. J.S. Sinninghe Damsté, M. Strous, W.I.C. Rijpstra, E.C. Hopmans, J.A.J. Geenevasen, A.C.T. Van Duin, L.A. Van Niftrik, M.S.M. Jetten, Linearly concatenated cyclobutane lipids form a dense bacterial membrane, Nature 419 (2002) 708–712.

100. W.R.L. van der Star, A.I. Miclea, U.G.J.M. van Dongen, G. Muyzer, C. Picioreanu, M.C.M. van Loosdrecht, The membrane bioreactor: a novel tool to grow anammox bacteria as free cells, Biotechnology and Bioengineering 101 (2008) 286–294.

101. K. Isaka, Y. Date, T. Sumino, S. Yoshie, S. Tsuneda, Growth characteristic of anaerobic ammonium-oxidizing bacteria in an anaerobic filtrated reactor, Applied Microbiology & Biotechnology 70 (2006) 47–52.

102. A. Dapena-Mora, I. Fernandez, J.L. Campos, A. Mosquera-Corral, R. Mendez, M.S.M. Jetten, Evaluation of activity and inhibition effects on Annamox process by batch tests based on the nitrogen gas production, Enzyme and Microbial Technology 40 (2007) 859–865.

103. M. Strous, J.G. Kuenen, M.S.M. Jetten, Key physiology of anaerobic ammonium oxidation, Applied & Environmental Microbiology 65 (1999) 3248–3250.

104. C. Fux, Biological nitrogen elimination of ammonium-rich sludge digester liquids, PhD Thesis, ETH-Zürich, Switzerland, 2003.

105. K. Egli, U. Fanger, P.J.J. Alvarez, H. Siegrist, J.R. Van Der Meer, A.J.B. Zehnder, Enrichment and characterization of an anammox bacterium from a rotating biological contactor treating ammonium-rich leachate, Archive of Microbiology 175 (2001) 198–207.

106. J. Dosta, I. Fernández, J.R. Vázquez-Padín, A. Mosquera-Corral, J.L. Campos, J. Mata-Álvarez, R. Méndez, Short- and

long-term effects of temperature on the Anammox process, Journal of Hazardous Materials 154 (2008) 688–693.

107. L. Dexiang, L. Xiaoming, Y. Qi, Z. Guangming, G. Liang, Y. Xiu, Effect of inorganic carbon on anaerobic ammonium oxidation enriched in sequencing batch reactor, Journal of Environmental Sciences 20 (2007) 940–944.

108. K. Pynaert, B.F. Smets, S. Wyffels, D. Beheydt, S.D. Siciliano, W. Verstraete, Characterization of an autotrophic nitrogen-removing biofilm from a highly loaded lab-scale rotating biological contactor, Applied & Environmental Microbiology 69 (2003) 3626–3635.

109. Z. Yang, S. Zhou, Y. Sun, Start-up of simultaneous removal of ammonium and sulfate in an anaerobic ammonium oxidation (anammox) process in an anaerobic upflow bio-reactor, Journal of Hazardous Materials 169 (2009) 113–118.

110. M. Strous, E. Van Gerven, J.G. Kuenen, M.S.M. Jetten, Effects of aerobic and microaerobic conditions on anaerobic ammonium-oxidizing (Anammox) sludge, Applied & Environmental Microbiology 63 (1997) 2446–2448.

111. M. Ruscalleda, H. López, R. Ganigué, S. Ouig, M. Balaguer, J. Colprim, Heterotrophic denitrification on granular anammox SBR treating urban landfill leachate, Water Science & Technology 58 (9) (2008) 1749–1755.

112. P.C. Sabumon, Anaerobic ammonia removal in presence of organic matter: a novel route, Journal of Hazardous Materials 149 (2007) 49–59.

113. C. Tang, P. Zheng, C. Wang, Q. Mahmood, Suppresion of anaerobic ammonium oxidizers under high organic content in high-rate Anammox UASB reactor, Bioresource Technology 101 (2010) 1762–1768.

114. N. Chamchoi, S. Nitisoravut, J.E. Schmidt, Inactivation of ANAMMOX communities under concurrent operation of anaerobic ammonium oxidation (ANAMMOX) and denitrification, Bioresource Technology 99 (2008) 3331–3336.

115. D. Güven, A. Dapena, B. Kartal, M.C. Schmid, B. Maas, K. van de PasSchoonen, S. Sözen, R. Mendez, H.J.M. Op den Camp, M.S.M. Jetten, M. Strous, I. Schmidt, Propionate oxidation by and methanol inhibition of anaerobic ammonium-oxidizing bacteria, Applied & Environmental Microbiology 71 (2005) 1066–1071.

116. W. Jianlong, K. Jing, The characteristics of anaerobic ammonium oxidation (ANAMMOX) by granular sludge from an EGSB reactor, Process Biochemistry 40 (2005) 1973–1978.

117. B.E. Rittmann, P.I. McCarty, Environmental Biotechnology: Principles and Applications, McGraw-Hill Companies, 2001.

118. M. Kumar, J. Lin, Co-existence of anammox and denitrification for simultaneous nitrogen and carbon removal—strategies and issues, Journal of Hazardous Materials (2010), doi:10.1016/j.jhazmat.2010.01.077.

119. D. Paredes, P. Kuschk, T.S.A. Mbwette, F. Stange, R.A. Müller, H. Köser, New aspects of microbial nitrogen transformations in the context of wastewater treatment—a review, Engineering Life Sciences 7 (1) (2007) 13–25.

120. B. Kartal, J. Rattray, L.A. van Niftrik, J. van de Vossenberg, M.C. Schmid, R.I. Webb, S. Schouten, J.A. Fuerst, J.S. Damsté, M.S.M. Jetten, M. Strous, Candidatus "Anammoxoglobus propionicus" a new propionate oxidizing species of anaerobic ammonium oxidizing bacteria, Systematic and Applied Microbiology 30 (1) (2007) 39–49.

121. Z. Quan, S. Rhee, J. Zuo, Y. Yang, J. Bae, R. Park, S. Lee, Y. Park, Diversity of ammonium-oxidizing bacteria in a granular sludge anaerobic ammoniumoxidizing (anammox) reactor, Environmental Microbiology 10 (11) (2008) 3130–3139.

122. T. Awata, T. Kindaichi, N. Ozaki, A. Ohashi, The activity of the Anammox in the presence of organic matter, in: International Conference on Civil and Environmental Engineering, ICCEE-2009/Oct. 2009/Pukyong National University, 2009.

123. S.K. Toh, N.J. Ashbolt, Adaptation of anaerobic ammonium-oxidising consortium to synthetic coke-ovens wastewater,

Applied Microbiology and Biotechnology 59 (2–3) (2002) 344–352.

124. S.K. Toh, R.I. Webb, N.J. Ashbot, Enrichment of autotrophic anaerobic ammonium-oxidizing consortia from various wastewaters, Microbial Ecology 43 (2002) 154–167.

125. B. Kartal, M. Koleva, R. Arsov, W. van der Star, M.S.M. Jetten, M. Strous, Adaptation of a freshwater anammox population to high salinity wastewater, Journal of Biotechnology 126 (2006) 546–553.

126. K. Windey, I. De Bo, W. Verstraete, Oxygen-limited autotrophic nitrification–denitrification (OLAND) in a rotating biological contactor treating high-salinity wastewater, Water Research 39 (2005) 4512–4520.

127. G. Cema, J. Wiszniowski, S. Zabczynski, E. Zablocka-Godlewska, A. Raszka, J. Surmacz-Górska, Biological nitrogen removal from landfill leachate by deammonification assisted by heterotrophic denitrification in a rotating biological contactor (RBC), Water Science & Technology 55 (8–9) (2007) 35–41.

128. K. Isaka, T. Sumino, S. Tsuneda, High nitrogen removal performance at moderately low temperature utilizing anaerobic ammonium oxidation reactions, Journal of Bioscience and Bioengineering 103 (5) (2006) 486–490.

129. B. Szatkowska, E. Plaza, Temperature as a factor influencing the Anammox process performance, Water and Environmental Management Series. Young Researchers 2006 (2006) 51–58.

130. S. Rysgaard, R.N. Glud, N. Risgaard-Petersen, D. Dalsgaard, Denitrification and anammox activity in arctic marine sediments, Limnology & Oceanography 49 (2004) 1493–1502.

131. K. Pynaert, B.F. Smets, D. Beheydt, W. Verstraete, Start-up of autotrophic nitrogen removal reactors via sequential biocatalyst addition, Environmental Science & Technology 38 (2004) 1228–1235.

132. M. Strous, Microbiology and application of anaerobic ammonium oxidation, PhD Thesis, TU Delft, 2000, 144 p.

133. T. Yamamoto, K. Takaki, T. Koyama, K. Furukawa, Long-term stability of partial nitritation of swine wastewater digester liquor and its subsequent treatment by Anammox, Bioresource Technology 99 (2008) 6419–6425.

134. C. Trigo, J.M. Campos, J.M. Garrido, R. Mendez, Start-up of the Anammox process in a membrane bioreactor, Journal of Biotechnology 126 (2006) 475–487.

135. B. Arrojo, A. Mosquera-Corral, J.L. Campos, R. Méndez, Effects of mechanical stress on Anammox granules in a sequencing batch reactor (SBR), Journal of Biotechnology 123 (2006) 453–463.

136. S. Wyffels, P. Boeckx, K. Pynaert, D. Zhang, O. Van Cleemput, G. Chen, W. Verstraete, Nitrogen removal from sludge reject water by a two-stage oxygenlimited autotrophic nitrification denitrification process, Water Science & Technology 49 (5–6) (2004) 57–64.

137. X. Hao, J.J. Heijnen, M.C.M. van Loosdrecht, Sensitivity analysis of a biofilm model describing a one-stage completely autotrophic nitrogen removal (CANON) process, Biotechnology and Bioengineering 77 (3) (2001) 266–277.

138. M. Nielsen, A. Bollmann, O. Sliekers,M. Jetten,M. Schmid,M. Strous, I. Schmidt, L.H. Larsen, L.P. Nielsen, N.P. Revsbech, Kinetics, diffusional limitation and microscale distribution of chemistry and organisms in a CANON reactor, FEMS Microbiology Ecology 51 (2) (2005) 247–256.

139. [139] P. Veys, H. Vandeweyer, W. Audenaert, A. Monballiu, P. Dejans, E. Jooken, A. Dumoulin, B.D. Meesschaert, S.W.H. Van Hulle, Performance analysis and optimization of autotrophic nitrogen removal in different reactor configurations: a modeling study, Environmental Technology, in press.

140. J.R. Vazquez-Padin, I. Figueroa, A. Mosquera-Corral, J.L. Campos, R. Méndez, Post-treatment of effluents from

anaerobic digesters by the Anammox process, Water Science & Technology 60 (2009) 1135–1143.

141. S. Lackner, A. Terada, B.F. Smets, Heterotrophic activity compromises autotrophic nitrogen removal in membrane-aerated biofilms: results of a modelling study, Water Research 42 (2008) 1102–1112.

142. Z. Gong, F. Yang, S. Liu, H. Bao, S. Hu, K. Furukawa, Feasibility of a membraneaerated biofilm reactor to achieve single-stage autotrophic nitrogen removal based on Anammox, Chemosphere 69 (2007) 776–784.

143. O.A. Sliekers, K. Third, W. Abma, J.G. Kuenen, M.S.M. Jetten, CANON and Anammox in a gas-lift reactor, FEMS Microbiology Letters 218 (2003) 339–344.

144. B. Szatkowska, G. Cema, E. Plaza, J. Trela, B. Hultman, One-stage system with partial nitritation and Anammox process in moving-bed biofilm reactor, Water Science and Technology (8–9) (2007) 19–26.

145. L. Kuai, W. Verstraete, Ammonium removal by the oxygen-limited autotrophic nitrification–denitrification system, Applied & Environmental Microbiology 64 (1998) 4500–4506.

146. K.A. Third, O. Sliekers, J.G. Kuenen, M.S.M. Jetten, The CANON system (Completely Autotrophic Nitrogen-removal Over Nitrite) under ammonium limitation: interaction and competition between three groups of bacteria, System & Applied Microbiology 24 (2001) 588–596.

147. A. Hippen, K.-H. Rosenwinkel, G. Baumgarten, C.F. Seyfried, Aerobic deammonification: a new experience in the treatment of wastewaters, Water Science & Technology 35 (10) (1997) 111–120.

148. B. Wett, Development and implementation of a robust deammonification process, Water Science & Technology 56 (7) (2007) 81–88.

149. C. Helmer, S. Kunst, S. Juretschko, M.C. Schmid, K.-H. Schleifer, M. Wagner, Nitrogen loss in a nitrifying biofilm system, Water Science & Technology 39 (7) (1999) 13–21.

150. C. Helmer-Madhok, M. Schmid, E. Filipov, T. Gaul, A. Hippen, K.-H. Rosenwinkel, C.F. Seyfried, M. Wagner, S. Kunst, Deammonification in biofilm systems: population structure and function, Water Science & Technology 46 (1–2) (2002) 223–231.

151. O.A. Sliekers, N. Derwort, J.L. Campos-Gomez, M. Strous, J.G. Kuenen, M.S.M. Jetten, Completely autotrophic nitrogen removal over nitrite in a single reactor, Water Research 36 (2002) 2475–2482.

152. H. De Clippeleir, S.E. Vlaeminck, M. Carballa, W. Verstraete, A low volumetric exchange ratio allows high autotrophic nitrogen removal in a sequencing batch reactor, Bioresource Technology 100 (2009) 5010–5015.

153. G. Sun, D. Austin, Completely autotrophic nitrogen-removal over nitrite in lab-scale constructed wetlands: evidence from a mass balance study, Chemosphere 68 (2007) 1120–1128.

154. Z. Dong, T. Sun, A potential new process for improving nitrogen removal in constructed wetlands—promoting coexistence of partial-nitrification and ANAMMOX, Ecological Engineering 31 (2007) 69–78.

155. A. Hippen, C. Helmer, S. Kunst, K.-H. Rosenwinkel, C.F. Seyfried, Six years' practical experience with aerobic/anoxic deammonification in biofilm systems, Water Science & Technology 44 (2–3) (2001) 39–48.

156. T. Gaul, E. Filipov, N. Schlösser, S. Kunst, C. Helmer-Madhok, Balancing of nitrogen conversion in deammonifying biofilms through batch tests and GC/MS, Water Science & Technology 46 (4–5) (2002) 157–162.

157. N. Jardin, A. Hippen, C.F. Seyfried, K.-H. Rosenwinkel, F. Greulich, Deammonifikation des Schlammwassers auf der Kläranlage Hattingen mit Hilfe des Schwebebettverfahrens, GWF 142 (2001) 479–484, in German.

158. A. Cornelius, K.-H. Rosenwinkel, Aerob/anoxische Deammonifikation stickstoffhaltiger Abwässer im KALDNES®-Biofilmverfahren, KA-Wasserwirtschaft, Abwasser, Abfall 49 (2002) 1398–1403, in German.

159. G. Innerebner, H. Insama, I.H. Franke-Whittle, B. Wett, Identification of anammox bacteria in a full-scale deammonification plant making use of anaerobic ammonia oxidation, Systematic and Applied Microbiology 30 (2007) 408–412.

160. A. Joss, D. Salzgeber, J. Eugster, R. Knig, K. Rottermann, S. Burger, P. Fabijan, S. Leumann, J.Mohn, H. Siegrist, Full-scale nitrogen removal from digester liquid with partial nitritation and anammox in one SBR, Environmental Science and Technology 43 (2009) 53.

161. S. Liu, Z. Gong, F. Yang, H. Zhang, L. Shi, K. Furukawa, Combined process of urea nitrogen removal in anaerobic Anammox co-culture reactor, Bioresource Technology 99 (2008) 1722–1728.

162. A. Gali, J. Dosta, M.C.M. van Loosdrecht, J. Mata-Alvarez, Two ways to achieve an anammox influent from real reject water treatment at lab-scale: Partial SBR nitrification and SHARON process, Process Biochemistry 42 (2007) 715–720.

163. Y. Xue, F. Yang, S. Liu, Z. Fu, The influence of controlling factors on the startup and operation for partial nitrification in membrane bioreactor, Bioresource Technology 100 (2009) 1055–1060.

164. C. Fux, V. Marchesi, I. Brunner, H. Siegrist, Anaerobic ammonium oxidation of ammonium-rich waste streams in fixed-bed reactors, Water Science & Technology 49 (11–12) (2004) 77–82.

165. K. Udert, C. Fux, M. Munster, T. Larsen, H. Siegrist, W. Gujer, Nitrification and autotrophic denitrification of source-separated urine, Water Science & Technology 48 (1) (2003) 119–130.

166. S. Wyffels, S.W.H. Van Hulle, P. Boeckx, E.I.P. Volcke, O. Van Cleemput, P.A. Vanrolleghem, W. Verstraete, Modelling and simulation of oxygen-limited partial nitritation in a membrane-assisted bioreactor (MBR), Biotechnology & Bioengineering 86 (2004) 531–542.

167. Y.-J. Feng, S.-K. Tseng, T.-H. Hsia, C.-M. Ho, W.-P. Chou, Partial nitrification of ammonium-rich wastewater as pretreatment for anaerobic ammonium oxidation (Anammox) using membrane aeration bioreactor, Journal of Bioscience and Bioengineering 104 (2007) 182–187.

168. R. Ganigue, H. Lopez, M.D. Balaguer, J. Colprim, Partial ammonium oxidation to nitrite of high ammonium content urban land fill leachates,Water Research 41 (2007) 3317–3326.

169. Z. Liang, J. Liu, Landfill leachate treatment with a novel process: Anaerobic ammonium oxidation (Anammox) combined with soil infiltration system, Journal of Hazardous Materials 151 (2008) 202–212.

170. S. Qiao, T. Yamamoto, M. Misaka, K. Isaka, T. Sumino, Z. Bhatti, K. Furukawa, High-rate nitrogen removal from livestock manure digester liquor by combined partial nitritation–anammox process, Biodegradation 21 (2010) 11–20.

171. M. Strous, J.G. Kuenen, J.A. Fuerst, M. Wagner, M.S.M. Jetten, The Anammox case—a new experimental manifesto for microbiological eco-physiology, Antonie van Leeuwenhoek 81 (2002) 693–702.

172. M. Strous, E. Van Gerven, Z. Ping, J.G. Kuenen, M.S.M. Jetten, Ammonium removal from concentrated waste streams with the Anaerobic Ammonium Oxidation (Anammox) process in different reactor configurations, Water Research 31 (1997) 1955–1962.

173. A. Dapena-Mora, J.L. Campos, A. Mosquera-Corral, M.S.M. Jetten, R. Mendez, Stability of the ANAMMOX process in a gas-lift reactor and a SBR, Journal of Biotechnology 110 (2004) 159–170.

174. I. Tsushima, Y. Ogasawara, T. Kindaichi, H. Satoh, S. Okabe, Development of high-rate anaerobic ammonium-oxidizing (anammox) biofilm reactors, Water Research 41 (8) (2007) 1623–1634.

175. W.R.L. van der Star, W.R. Abma, D. Blommers, J.W. Mulder, T. Tokutomi, M. Strous, C. Picioreanu, M.C.M. Van Loosdrecht, Startup of reactors for anoxic ammonium oxidation: experiences from the first full-scale anammox reactor in Rotterdam, Water Research 41 (2007) 4149–4163.

176. T. Wang, H. Zhang, F. Yang, S. Liu, Z. Fu, H. Chen, Start-up of the Anammox process from the conventional activated sludge in a membrane bioreactor, Bioresource Technology 100 (2009) 2501–2506.

177. L. Zhang, P. Zheng, C. Tang, R. Jin, Anaerobic ammonium oxidation for treatment of ammonium-rich wastewaters, Journal of Zhejiang University Science 9 (5) (2008) 416–426.

178. H. Park, A. Rosenthal, A. Deur, K. Beckmann, K. Ramalingam, J. Fillos, K. Chandran, Molecular Based Characterisation of the Microbial Ecology and Activity of Anammox Bioreactors, Water Environment Federation, WEFTEC, 2009.

179. K.-I. Gil, E. Choi, Nitrogen removal by recycle water nitritation as an attractive alternative for retrofit technologies in municipal wastewater treatment plants, Water Science & Technology 49 (5–6) (2004) 39–46.

180. P. Jenicek, P. Svehla, J. Zabranska, M. Dohanyos, Factors affecting nitrogen removal by nitritation/denitritation, Water Science & Technology 49 (5–6) (2004) 73–79.

181. S. Wyffels, P. Boeckx, K. Pynaert, W. Verstraete, O. Van Cleemput, Sustained nitrite accumulation in a membrane-assisted bioreactor (MBR) for the treatment of ammonium rich wastewater, Journal of Chemical Technology & Biotechnology 78 (2003) 412–419.

182. M. Chen, J.-H. Kim, N. Kishida, O. Nishimura, R. Sudo, Enhanced nitrogen removal using C/N load adjustment and real-time control strategy in sequencing batch reactors for

swine wastewater treatment, Water Science & Technology 49 (5–6) (2004) 309–314.

183. D. Obaja, S. Macé, J. Costa, C. Sans, J. Mata-Alvarez, Nitrification, denitrification and biological phosphorus removal in piggery wastewater using a sequencing batch reactor, Bioresource Technology 87 (2003) 103–111.

184. K. Poo, B. Jun, S. Lee, H. Woo, C. Kim, Treatment of strong nitrogen swine wastewater at full-scale sequencing batch reactor, Water Science & Technology 49 (5–6) (2004) 315–323.

185. A. Tilche, E. Bacilieri, G. Bortone, F. Malaspina, S. Piccinini, L. Stante, Biological phosphorus and nitrogen removal in a full scale sequencing batch reactor treating piggery wastewater, Water Science & Technology 40 (1) (1999) 199–206.

186. J. Chung, W. Bae, Y. Lee, G. Ko, S. Lee, S. Park, Investigation of the effect of free ammonia concentration upon leachate treatment by shortcut biological nitrogen removal process, in: IWA Speciality Symposium on Strong Nitrogenous and Agro-Wastewater, vol. 1, Seoul, Korea, June 11–13, 2003, pp. 93–104.

187. J. Jokela, R. Kettunen, K. Sormunen, J. Rintal, Biological nitrogen removal from municipal landfill leachate: low-cost nitrification in biofilters and laboratory scale in-situ denitrification, Water Research 36 (2002) 4079–4087.

188. S. Kalyuzhnyi, M. Gladchenko, Sequenced anaerobic-aerobic treatment of high strength, strong nitrogenous landfill leachates, Water Science & Technology 49 (5–6) (2004) 301–312.

189. U. Abeling, C.F. Seyfried, Anaerobic aerobic treatment of potato-starch wastewater, Water Science & Technology 28 (2) (1993) 165–176.

190. U. Austermann-Haun, H. Meyer, C. Seyfried, K.-H. Rosenwinkel, Full scale experiences with anaerobic/aerobic treatment plants in the food and beverage industry, Water Science & Technology 40 (1) (1999) 305–312.

191. P. Deng Petersen, K. Jensen, P. Lyngsie, N. Hendrik Johansen, Nitrogen removal in industrial wastewater by nitration and denitration—3 years of experience, Water Science & Technology 47 (11) (2003) 181–188.

192. B. Kartal, L. van Niftrik, O. Sliekers, M.C. Schmid, I. Schmidt, K. van de PasSchoonen, I. Cirpus, W. van der Star, M. van Loosdrecht, W. Abma, J.G. Kuenen, J. Mulder, M.S.M. Jetten, H. op den Camp, M. Strous, J. van de Vossenberg, Application, eco-physiology and biodiversity of anaerobic ammonium oxidizing bacteria, Reviews in Environmental Science and Biotechnology 3 (2004) 255–264.

193. M. Schmid, U. Twachtmann, M. Klein, M. Strous, S. Juretschko, M. Jetten, J.W. Metzger, K.H. Schleifer, M. Wagner, Molecular evidence for genus level diversity of bacteria capable of catalyzing anaerobic ammonium oxidation, Systematic and Applied Microbiology 23 (2000) 93–106.

194. D. Woebken, P. Lam, M.M.M. Kuypers, S.W.A. Naqvi, B. Kartal, M. Strous, M.S.M. Jetten, B.M. Fuchs, R. Amann, A microdiversity study of anammox bacteria reveals a novel Candidatus Scalindua phylotype in marine oxygen minimum zones, Environmental Microbiology 10 (11) (2008) 3106–3119.

195. B. Wett, Solved upscaling problems for implementing deammonification of rejection water, Water Science & Technology 53 (2006) 121–128.

196. G. Nyhuis, V. Stadler, B. Wett, Inbetriebnahme der ersten DEMON-Anlage in der Schweiz zur direkten Stickstoffelimination und erste Betriebsergebnisse (Taking into operation the First DEMON installation in Switzerland for direct nitrogen removal and first process results) 6, Aachener Tagung mit Informationsforum, Stickstoffrückbelastung – Stand der Technik 2006, Aachen Germany, ATEMIS GmbH, 2006.

197. S.E. Vlaeminck, L.F.F. Cloetens, H. De Clippeleir, M. Carballa, W. Verstraete, Granular biomass capable of partial nitration

and anammox, Water Science & Technology 59 (3) (2009) 609–617.

198. M. Denecke, V. Rekers, U. Walter, Einsparpotenziale bei der biologischen Reinigung von Deponiesickerwasser (Cost saving potentials in the biological treatment of landfill leachate), Muell und Abfall 39 (2007) 4–7.

199. V. Rekers, M. Denecke, U.Walter, Betriebserfahrung mit der anaeroben Deammonifikation von Deponiesickerwasser (Operational experiences with the anaerobic deammonification of landfill leachate), Depotech, 2008, Leoben, Austria, Institut für nachhaltige Abfallwirtschaft und Entsorgungstechnik, 2008.

200. P. Johansson, A. Nyberg, M. Beier, A. Hippen, C.F. Seyfried, K.-H. Rosenwinkel, 1998, Cost efficient sludge liquor treatment, Royal Institute of Technology, Stockholm, TRITA-AMI Report 3048.

201. M. Beier, Y. Schneider, Abschlussbericht Entwickelung von Bilanzmodellen für die Prozesse Deammonifikation und Nitritation zur Abbildung gross- and halb-technischer Anlagen (Final report development of balance models for deammonification and nitritation processes to illustrate full and half technical scale installations), Leibniz University Hannover, Hannover, 2008.

202. A. Rosenthal, K. Ramalingam, H. Park, A. Deur, K. Beckmann, K. Chandran, J. Fillos, Anammox studies using New York city centrate to correlate performance, population dynamics and impact of toxins, in: WEFTEC 2009, 2009.

203. R.-C. Jin, P. Zheng, A.-H. Hu, Q. Mahmood, B.-L. Hu, G. Jilani, Performance comparison of two anammox reactors: SBR and UBF, Chemical Engineering Journal 138 (2008) 224–230.

204. A. Dapena-Mora, S.W.H. Van Hulle, J.L. Campos, R. Mendez, P.A. Vanrolleghem, M.S.M. Jetten, Enrichment of Anammox biomass from municipal activated sludge: experimental and modelling results, Journal of Chemical Technology & Biotechnology 79 (2004) 1421–1428.

205. K.A. Third, J. Paxman, M. Schmid, M. Strous, M.S.M. Jetten, R.C. Cord-Ruwisch, Enrichment of Anammox from Activated Sludge and Its Application in the CANON Process, Microbial Ecology 49 (2005) 236–244.

206. J. Vazquez-Padin, I. Fernádez, M. Figueroa, A. Mosquera-Corral, J. Campos, R. Méndez, Applications of Anammox based processes to treat anaerobic digester supernatant at room temperature, Bioresource Technology 100 (2009) 2988–2994.

207. U. Imajo, T. Tokutomi, K. Furukawa, Granulation of Anammoxmicroorganisms in up-flow reactors, Water Science & Technology 49 (5–6) (2004) 155–164.

208. J.R. Vazquez-Padin, M.J. Pozo, M. Jarpa, M. Figueroa, A. Franco, A. MosqueraCorral, J.L. Campos, R. Méndez, Treatment of anaerobic sludge digester effluents by the CANON process in an air pulsing SBR, Journal of Hazardous Materials 166 (2009) 336–341.

Chapter 4

Targeting Minimum Waste Treatment Flow Rate

Santanu Bandyopadhyay

Energy Science and Engineering, Indian Institute of Technology Bombay, Powai, Mumbai 400076, India

ABSTRACT

Due to various environmental regulations, it is important to minimize the cost associated with treatment of different industrial wastes prior to its discharge to the environment. In this paper, an algebraic methodology, based on the principles of process integration, is proposed to target the minimum waste treatment flow rate to satisfy environmentally safe discharge limit. An associated graphical representation of the optimization problem is also provided to gain physical insight. In the proposed methodology, the treatment units are modelled either as unit with constant outlet concentration or as a unit with fixed removal ratio. There is flow loss associated with the treatment unit. The flow loss is assumed to be proportional to the inlet flow rate. Applicability of the proposed methodology is demonstrated through examples from water management, volatile organic compound treatment and flue gas desulphurization.

INTRODUCTION

The cost of treating different waste streams is increasing steadily as environmental regulations are becoming more and more stringent. It is, therefore, important to minimize the cost associated with treatment of these wastes while satisfying the environmental norms. Targeting and designing waste allocation network for optimal treatment can be effectively addressed through process integration techniques. Techniques of process integration are primarily used for process design (both grassroots and retrofits) with special emphasis on efficient utilization of resources and reducing environmental pollution. Techniques of process integration were originally developed for analyzing heat exchanger networks[1] and [2] and integration of energy intensive equipments [3] and [4]. Later, these techniques are evolved to address mass exchanger networks [5] and [6] and water networks [7], [8] and [9]. Recently, the techniques of process integration have been applied for design and optimization of various energy systems[10], [11], [12], [13], [14], [15], [16] and [17]. Bandyopadhyay [18] demonstrated the applicability of targeting tools for waste reduction. Primary objective of this paper is to propose an algebraic methodology, based on the principles of process integration, for targeting the minimum waste treatment flow rate to satisfy environmentally safe discharge limit.

Techniques of process integration may be classified into two broad categories: graphical pinch analysis-based approaches and approaches based on mathematical optimization techniques.

Graphical pinch analysis-based approaches help in getting a physical insight of the problem through its graphical representations and simplified tableau-based calculation procedures. On the other hand, mathematical optimization-based methodologies are preferred to address issues like multiple contaminants, controllability, flexibility, cost-optimality, etc. In this paper, an algebraic approach based on tableau-based calculation procedure is proposed and associated graphical representation is also provided to gain physical insight of the problem. However, the proposed methodology is restricted to single contaminant only.

In a seminal paper, Takama et al. [19] solved the complete water management problem using superstructure-based non-linear optimization technique. Wang and Smith [20] have developed a systematic approach for designing distributed effluent treatment systems. This procedure has been extended by Kuo and Smith [21] for multiple treatment processes. Mathematical optimization techniques have also been used to design distributed effluent treatment systems [22] and [23]. Freitas et al. [24]proposed the use of the hierarchical design approach [25], supplemented with a database and expert system to determine the best sequence of treatment processes. However, such a method cannot guarantee the optimality. Statyukha et al. [26] proposed a hybrid approach for designing wastewater treatment networks. The insight-based technique is employed to obtain an initial solution and then superstructure-based non-linear optimization is solved. Zhelev and Bhaw [27] introduced combined water and oxygen pinch analysis for designing optimum wastewater treatment network. The minimum oxygen requirement for waste degradation was targeted in combination of the water pinch analysis. Interaction between operations that use water and effluent treatment systems have also been addressed [8], [28], [29],[30], [31], [32] and [33]. Alva-Argáez et al. [34] and [35] addressed the entire water management problem through superstructure-based MINLP formulation.

In literature, treatment units are either modelled as unit with constant outlet concentration (e.g., filtration, and membrane separation systems) or as a unit with fixed removal ratio (e.g., scrubber). These units are typically modelled without any flow loss. Most treatment units such as membrane separation systems (e.g., microfiltration, ultrafiltration, reverse osmosis, etc.), flotation systems (e.g., dissolved air flotation, induced air flotation, etc.), gravity and settling systems (e.g., coagulation, flocculation, clarification, etc.), filtration systems (e.g., granular bed, vacuum drum, press, belt filter, etc.), etc. have flow loss associated with them. For example, a membrane-based treatment system separates a feed stream into two product streams; a lower concentration permeate and a higher concentration retentate (or reject) streams. In literature such treatment units with two product streams are termed as partitioning treatment units [36] and [37]. Typically, higher concentrate reject stream form a partitioning unit is not reused in the process and is sent for treatment separately. Mathematical optimization-based methodology has been proposed in the context of water regeneration and reuse [36]. In this paper, treatment units are assumed to have flow loss and applied to satisfy the environmental discharge norms. An algebraic targeting methodology is developed to incorporate such treatment units with flow loss in the process design to satisfy the environmental regulations. Applicability of the proposed methodology is demonstrated through different illustrative examples.

PROBLEM STATEMENT

The general problem of targeting the minimum waste treatment flow rate using waste composite curve may be mathematically stated as follows. In a process, a set of Nw waste sources is given. Each waste source produces a flow F_{wi} with a given contaminant concentration of C_{wi}. As the environmental regulations imposed on the overall plant, C_e denotes the concentration below which waste may be discharged to the environment. These waste streams have to be treated and contaminant has to be removed in the treatment plant. The objective of this work is to develop an algorithmic technique with graphical representation to target the minimum effluent flow rate to be treated in the treatment plant. In this paper, we assume that only one treatment plant is sufficient for the purpose and the proposed methodology is restricted to single contaminant.

An effluent treatment unit with flow loss may be modelled in two ways: constant outlet concentration and constant removal factor. Furthermore, there may be an added constraint on the maximum inlet concentration to a treatment unit. For both kinds of treatment units, the outlet flow rate from the treatment unit (F_{Tout}) is assumed to be proportional to the inlet flow rate to the treatment unit (F_{Tin}).

$$F_{Tout} = \alpha F_{Tin} \tag{1}$$

Eq. (1) suggests that $(1 - \alpha)$ times the inlet flow rate to the treatment unit (F_{Tin}) is lost during treatment and α may be called the flow factor of the treatment unit. It may be noted that treatment units are not modelled as partitioning treatment units [36] and [37] in this paper. A partitioning treatment unit produces a highly concentrated reject stream that has to be treated separately in another treatment unit. In this paper, treatment units are simply modelled with flow loss and it is assumed that the lost flow cannot be recovered as a separate stream. However, a partitioning treatment unit can also be modelled as a treatment unit with flow loss and higher concentration reject stream has to be treated separately or may be converted into by-product. Treatment of such reject streams or production of saleable by-products are beyond the scope of the proposed algebraic methodology.

For a treatment unit with constant outlet concentration (C_{Tout}), the treated waste comes out of the treatment unit at a fixed concentration of C_{Tout} and this is independent of the inlet concentration (C_{Tin}). On the other hand, for a treatment unit with constant removal factor, the outlet concentration (C_{Tout}) of the treated effluent depends on the inlet concentration of the effluent to the treatment unit (C_{Tin}). The removal ratio (r) of the treatment unit is defined as

$$r = \frac{F_{Tin} C_{Tin} - F_{Tout} C_{Tout}}{F_{Tin} C_{Tin}} \tag{2}$$

Additional constraint in the form of the maximum inlet concentration to the treatment unit may also be imposed for overall optimization.

The objective is to minimize the inlet flow rate to the waste treatment plant (F_{Tin}) such that the concentration of the remaining waste, after mixing with the treated one, is less than the specified environmental discharge limit (C_e).

TARGETING MINIMUM EFFLUENT FLOW RATE

Bandyopadhyay et al. [8] have proposed a novel limiting composite curve, called the source composite curve, to simultaneously target the minimum freshwater requirement, the maximum water reuse, the minimum wastewater generation, and the minimum effluent to be treated to meet environmental norms. To target the minimum effluent to be treated to meet the environmental regulation, a wastewater composite curve was proposed [8]. It may be noted that the wastewater composite curve is equivalent to the original source composite curve without any internal demand. The wastewater composite curve is plotted on contaminant load (M) vs. concentration (C) diagram. Generation of the waste composite curve (equivalent to the wastewater composite curve) for a given set of waste sources is discussed briefly before developing the targeting methodologies. Formulae for each step are tabulated in Table 1.

Step 1:

Concentrations of all waste sources including the environmental limit are tabulated in decreasing order in the first column. If value of a particular concentration occurs more than once, the same need not be repeated. Without loss of generality, it can be said that the concentration for kth row is denoted as Ck such that

$$C_1 > C_2 > \cdots > Ck > \cdots > Cn$$

(3)

In may be noted that the last entry of this column should be zero $(C_n = 0)$.

Step 2:

Net waste flows (i.e., sum of waste flow rates corresponding to a particular concentration) are tabulated in second column. For kth row, net flow rate is denoted as F_k (Table 1).

Step 3:

Cumulative waste flow rates are tabulated in the third column. Summation of net waste flow rates for all previous rows $(\sum_{l=1}^{k} F_l)$ denotes the cumulative flows for kth row. Last entry in this column suggests the total waste available (F_T) for a given problem.

Step 4:

Fourth column represents the contaminant load (m_k) for each concentration interval. Contaminant load is defined as the product of the concentration with the flow rate. First entry in fourth column is assigned to be 0. For all subsequent rows, the difference between the last two concentrations is multiplied by the cumulative flow rates, tabulated in third column, to calculate the concentration load. Mathematically, concentration load (m_k) for each concentration interval can be calculated using the following formula.

$$m_k \ = 0 \qquad\qquad\qquad \text{for } k = 1$$
$$= (C_{k-1} - C_k)(\sum_{l=1}^{k-1} F_l) \quad \text{for } k > 1 \qquad (4)$$

Step 5:

Cumulative contaminant loads are calculated by summing contaminant loads for all previous rows$(M_k = \sum_{l \le k} ml)$ and tabulated in the fifth column. Using Eq. (4), cumulative contaminant load for kth row may be expressed as

$$M_k = 0 \qquad\qquad \text{for } k = 1$$
$$= \sum_{l=1}^{k-1} F_l(C_l - C_k) \quad \text{for } k > 1 \tag{5}$$

Last entry in this column suggests the total contaminant load available (M_T) for a given problem.

$$M_T = \sum_{l=1}^{n} m_l = \sum_{l=1}^{n-1} F_l C_l \tag{6}$$

Table 1: Tabular representation for generation of waste composite curve

	First column	Second column	Third column	Fourth column	Fifth column
	Concentration	Net waste flows	Cumulative flows	Contaminant load	Cumulative contaminant load
First row	C_1	F_1	F_1	$m_1 = 0$	$M_1 = m_1 = 0$
Second row	C_2	F_2	$F_1 + F_2$	$m_2 = F_1(C_1 - C_2)$	$M_2 = m_1 + m_2 = F_1(C_1 - C_2)$
...	...	...	...	...	...
kth row	C_k	F_k	$\sum_{l=1}^{k} F$	$m_k = (C_{k-1})(\sum_{l=1}^{k-1} F_l)$	$m_k = \sum_{l=1}^{k} m_l = \sum_{l=1}^{k-1} F_l(C_l - C_k)$
...	...	...	...	...	...
nth (last) row	$C_n = 0$	F_n	$\sum_{l=1}^{k} F_l = 1$	$m_n = C_{n-1}(\sum_{l=1}^{n-1} F_l)$	$m_T = \sum_{l=1}^{n} m_l = \sum_{l=1}^{n-1} F_l C_l$

Now fifth column (cumulative contaminant load) may be plotted against the first column (concentration) to obtain the waste composite curve.

Treatment Unit with Constant Outlet Concentration

The outlet concentration of the waste, treated in a treatment unit with constant outlet concentration, is always fixed at C_{Tout}. The relation between the outlet flow rate ($F_{To}ut$) from the treatment unit and the inlet flow rate (F_{Tin}) to the treatment unit is given by Eq. (1). Total contaminant load removed (M_R) by the treatment unit is expressed as follows:

$$MR = F_{Tin}C_{Tin} - F_{Tout}C_{Tout} = F_{Tin}(C_{Tin} - \alpha C_{Tout}) \tag{7}$$

Rearranging the above equation, the inlet concentration to the treatment unit can be expressed as follows:

$$C_{Tin} = \frac{M_R}{F_{Tin}} + \alpha C_{Tout} \tag{8}$$

For targeting the minimum effluent flow rate to be treated in the treatment without any flow loss ($\alpha = 1$), treatment line is rotated on the contaminant load (M) vs. concentration (C) diagram. The minimum treatment flow rate is targeted by rotating the treatment line with point (M_R, C_{Tout}) as the pivot point such that it just touches the waste composite curve. The point at which the treatment line touches the waste composite curve represents the treatment pinch point and the point at which it touches the concentration axis represents the inlet concentration to the treatment unit. However, the same methodology cannot be applied directly due to the flow loss associated with the treatment unit. For a treatment unit with constant outlet concentration and flow loss, the treatment line on a contaminant load (M) vs. concentration (C) diagram must pass through the points $(0, C_{Tin})$ and $(F_{Tin}\{C_{Tin} - C_{Tout}\}, C_{Tout})$. Using Eq. (8), $F_{Tin}(C_{Tin} - C_{Tout})$ can be simplified as $M_R - (1 - \alpha) F_{Tin}C_{Tout}$. Therefore, the treatment line must pass through the points $(0, C_{Tin})$ and $(MR - \{1 - \alpha\} F_{Tin}C_{Tout}, C_{Tout})$. If the point (M_k, C_k) on the source composite curve holds the treatment pinch, the effluent flow rate at the inlet of the treatment unit is calculated to be:

$$f_{Tk} = \frac{M_R - M_k}{C_k - \alpha C_{Tout}}$$

$$(9)$$

After treating a portion of the waste in the treatment unit, remaining waste, including the treated one, may be discharged to the environment. Therefore, a total of $F_T - (1 - \alpha) f_{Tk}$ waste is discharged with a contaminant load of $M_T - M_R$. To satisfy the environmental discharge limit, following equation must be satisfied.

$$\frac{M_T - M_R}{F_T - (1 - \alpha) f_{Tk}} \leq C_e$$

$$(10)$$

Eliminating M_R form Eqs. (9) and (10), the treatment flow rate can be expressed as

$$f_{Tk} \geq \frac{M_T - F_T C_e - M_k}{C_k - \alpha C_{Tout} - (1 - \alpha) C_e}$$

$$(11)$$

The above inequality can be applied to target the minimum waste flow rate to be treated in the treatment unit. Algorithmic step, in continuation of the previous steps, is described below to target the minimum waste treatment flow rate.

Step 6: Waste flow rates corresponding to the contaminant concentrations (tabulated in the first column) and the cumulative contaminant loads (tabulated in the fifth column), are calculated applying the following equation and tabulated in the sixth column.

$$\begin{aligned} f_{Tk} &= \frac{M_T - F_T C_e - M_k}{C_k - \alpha C_{Tout} - (1 - \alpha) C_e} \quad \text{when } C_k \geq C_{Tout} \\ &= 0 \quad \text{otherwise} \end{aligned}$$

$$(12)$$

Maximum entry in this column defines the minimum waste flow rate to be treated in the treatment unit.

For some treatment unit, an additional constraint in the form of the maximum allowable inlet concentration ($C_{Tin\,max}$) to the treatment unit is also specified. The minimum waste flow rate to the treatment unit that satisfy the additional constraint may be calculated directly using the following equation:

$$f_T = \frac{M_T - F_T C_e}{C_{Tin\,max} - \alpha C_{Tout} - (1 - \alpha) C_e}$$

$$(13)$$

The minimum waste treatment flow rate to be treated in the treatment unit is the maximum of the flow rates obtained using Eqs. (12) and (13).

Treatment Unit with Constant Removal Ratio

For a treatment unit with constant removal factor, the relation between the outlet concentration (C_{Tout}) of the treated waste and the inlet concentration (C_{Tin}) to the treatment unit is given by Eq. (2). Combining Eq.(1) with the definition of the removal ratio, the inlet concentration to the treatment unit can be expressed as a function of the outlet concentration.

$$C_{Tin} = \frac{\alpha}{1-r} C_{Tout}$$

(14)

Total contaminant load removed (M_R) by the treatment unit is simplified as follows:

$$MR = F_{Tin}C_{Tin} - F_{Tout}C_{Tout} = F_{Tin}C_{Tin}r$$

(15)

For targeting the minimum waste flow rate to be treated in the treatment unit without any flow loss ($\alpha = 1$), treatment line is rotated on the contaminant load (M) vs. concentration (C) diagram. The pivot point for targeting the minimum treatment flow rate is $(M_R/r, 0)$. It may be noted that the pivot points are different for different types of treatment units. Similar to the previous section, the point at which the treatment line touches the waste composite curve represents the treatment pinch point and the point at which it touches the concentration axis represents the inlet concentration to the treatment unit. Due to flow loss associated with the treatment unit, the same methodology cannot be applied directly.

Similar to the previous section, the treatment line on a contaminant load (M) vs. concentration (C) diagram must pass through the points $(0, C_{Tin})$ and $(F_{Tin} \{C_{Tin} - C_{Tout}\}, C_{Tout})$. Using Eqs. (14) and (15), $F_{Tin}(C_{Tin} - C_{Tout})$ can be simplified as $M_R (1 - \alpha - r)/r\alpha$. Therefore, the treatment line must pass through the points $(0, C_{Tin})$

and $(M_R (1 - \alpha - r)/r\alpha, C_{Tout})$. The waste flow rate at the inlet of the treatment unit, if the point (M_k, C_k) on the waste composite curve holds the treatment pinch, is calculated to be:

$$f_{Tk} = \frac{M_R/r - M_k}{C_k} \tag{16}$$

Similar to the previous section, after treating a portion of the waste in the treatment unit, remaining waste, including the treated one, may be discharged to the environment. To satisfy the environmental discharge limit, Eq. (10) has to be satisfied. Eliminating M_R from Eqs. (10) and (16), the treatment flow rate can be expressed as

$$f_{Tk} \geq \frac{M_T - F_T C_e - r M_k}{r C_k - (1 - \alpha) C_e} \tag{17}$$

Similar to Eq. (11), Eq. (17) can be applied to target the minimum waste flow rate to be treated in the treatment unit having constant removal ratio. An algorithmic step is described below to target the minimum waste treatment flow rate.

Step 6: Waste flows corresponding to the contaminant concentrations (tabulated in the first column) and the cumulative contaminant loads (tabulated in the fifth column), are calculated applying the following equation and tabulated in the sixth column.

$$
\begin{aligned}
f_{Tk} &= \frac{M_T - F_T C_e - r M_k}{r C_k - (1 - \alpha) C_e} \quad \text{when } M_k \leq M_c \\
&= 0 \quad \text{otherwise}
\end{aligned} \tag{18}
$$

where the critical contaminant mass load (M_c) is calculated as follows:

$$M_c = \frac{(M_T - F_T C_e)(\alpha + r - 1)}{r\alpha - (1 - \alpha)(1 - r)(C_e/C_k)} \tag{19}$$

Eq. (19) is equivalent to the condition that $C_k \geq C_{Tout}$. Maximum entry in this column defines the minimum waste flows to be treated in the treatment unit.

In absence of the additional constraint of the maximum allowable inlet concentration $(C_{Tin\ max})$ to the treatment unit, Eq. (18) targets the minimum waste flow rate to be treated in the treatment unit. Similar to Eq. (13), the minimum waste flow rate to the treatment unit that satisfy the additional constraint of the maximum allowable inlet concentration ($C_{Tin\,max}$) to the treatment unit, may be calculated directly using the following equation:

$$f_T = \frac{M_T - F_T C_e}{r C_{Tin\,max} - (1 - \alpha) C_e}$$

(20)

The minimum waste treatment flow rate to be treated in the treatment unit is the maximum of the flow rates obtained using Eqs. (18) and (20).

For a low value of both removal ratio and flow factor, the environmental discharge concentration controls the pinch point. In such cases, treated waste from the treatment unit has to be recycled across the treatment unit. However, due to flow loss associate with the treatment unit, there exists a minimum value of flow factor for which the targeting step is physically meaningful. The minimum value of the flow factor is expressed as follows:

$$\alpha_{min} = \frac{M_T(1 - r)}{M_T(1 - r) - r F_T C_e}$$

(21)

The limiting flow rate to the treatment unit is given as

$$f_{T\,max} = F_T \frac{M_T(1 - r)}{r C_e}$$

(22)

If the actual alpha is lower than the minimum, the entire waste to the treatment unit is lost and no waste is produced. It suggests that such a simple model cannot be used and a more realistic model has to be used.

Application of the proposed algorithm is demonstrated through the following illustrative examples.

ILLUSTRATIVE EXAMPLES

In this section different illustrative examples are considered to demonstrate the applicability of the proposed methodology. Examples are considered from the field of water management, treatment of volatile organic compounds (VOCs), and desulphurization of flue gases.

Example 1: Water Treatment through a Treatment Unit with Constant Outlet Concentration

This example is taken from the field of wastewater treatment. The wastewater and treatment unit data for this example are given in Table 2. There are two wastewater sources. The outlet concentration of the treatment unit is fixed at 25 ppm. In the first case, no restriction related to the inlet concentration to the treatment unit is imposed. Later on, an additional constraint on the maximum allowable inlet concentration of 500 ppm to the treatment unit is imposed. The objective is to target the minimum amount of effluent to be treated in the treatment unit to satisfy the environmental discharge limit of 30 ppm.

Table 2: Wastewater and treatment unit data for example 1

#	Contaminant concentration (ppm)	Flow rate (t/h)
1	800	50
2	400	100

Environmental limit for discharge concentration, C_e = 30 ppm. Characteristics of the treatment unit, = 0.8, C_{Tout} = 25 ppm, and $C_{Tin\ max}$ = 500 ppm.

The steps of the proposed algorithm are shown in Table 3. In the first column of Table 3, concentrations of all water sources, including the environmental limit, are tabulated in decreasing order (step 1). Net waste flows, corresponding to each concentration in column one, is tabulated in the second column of Table 3 (step 2). Cumulative waste flow rates, as described in step 3 of the proposed algorithm, are tabulated in the third column. Contaminant mass loads, calculated using Eq. (4), are tabulated in the forth column of Table 3. Cumulative contaminant mass loads, calculated using step 5 of the proposed algorithm, are tabulated in column five of Table 3. For this example, the outlet concentration of the treatment unit is specified to be 25 ppm. Applying Eq. (12), treatment flow rates, corresponding to different contaminant concentrations, are calculated and tabulated in the sixth column of Table 3 (step 6).

Table 3: Generation of waste composite curve and targeting for minimum effluent flow rate to be treated for example 1

Contaminant concentration (ppm)	Net flow rate (t/h)	Cumulative flow rate (t/h)	Contaminant mass load (kg/h)	Cumulative mass load (kg/h)	Treatment flow rate (t/h)
800	50	50	0	0	97.55
400	100	150	20	20	148.40
30	0	150	55.5	75.5	0
25	0	150	0.75	76.25	0
0	0	150	3.75	80	0

The sixth column of Table 3 suggests that 148.4 t/h of effluent to be treated in the treatment unit to satisfy the environmental norm. Waste composite curve, shown in Fig. 1, is obtained by plotting fifth column against the first column. The treatment line is also shown in Fig. 1. Form Table 3 as well as from Fig. 1, the treatment pinch is identified to be 400 ppm. According to the pinch principles [20], any wastewater with concentration higher than the pinch concentration has to be treated in the treatment unit. Therefore, 50 t/h of wastewater at 800 ppm and 98.4 t/h of wastewater at 400 ppm must be put to the treatment unit. The inlet concentration of the wastewater to the treatment unit is 534.8 ppm. Remaining 1.6 t/h of wastewater at 400 ppm bypasses the treatment unit and mixed with the treated water to satisfy the discharge limit of 30 ppm. The wastewater allocation network is shown in Fig. 2a.

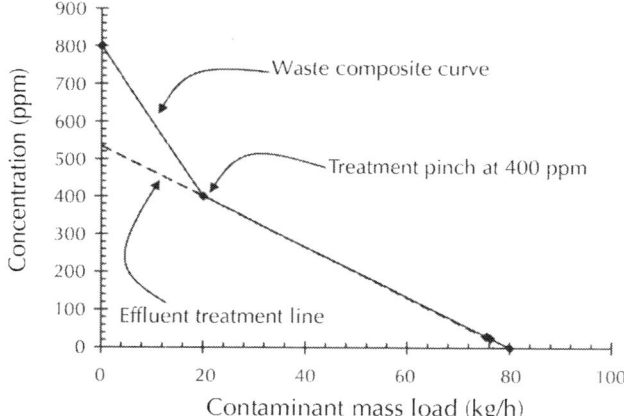

Figure 1: Waste composite curve for example 1 and effluent treatment line for unconstraint case.

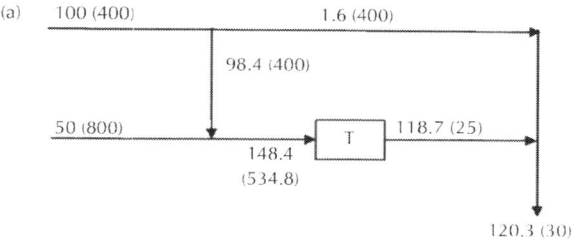

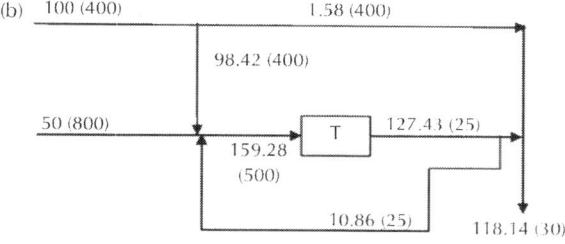

Figure 2: Wastewater allocation networks for example 1: (a) without any restriction on the inlet concentration to the treatment unit, and (b) the maximum allowable inlet concentration to the treatment unit is restricted

to 500 ppm. (The values show flow rate in t/h with contaminant concentrations in ppm within parenthesis.).

Without considering the flow loss associated with the treatment unit, the target for the minimum effluent flow rate to be treated in the treatment unit would have been 148 t/h. Due to flow loss associate with the treatment unit, the minimum flow rate of the effluent to be treated in the treatment unit is increased by 0.27% and the total wastewater discharged to the environment is reduced by 19.8%.

If the maximum allowable inlet concentration to the treatment unit is restricted to 500 ppm, Eq. (13) may be used to target the minimum effluent to be treated in the treatment unit. Treatment flow rate of 159.28 t/h satisfies the environmental limit. The treatment line for this case is not shown for brevity. Due to restriction on the maximum allowable inlet concentration to the treatment unit, the minimum flow rate of the effluent to be treated in the treatment unit is increased by 7.3% and the total wastewater discharged to the environment is reduced by 1.8%. It may be noted that there is no treatment pinch for this example. It may also be noted that the minimum effluent flow rate to be treated in the treatment unit is more than the total wastewater available. To satisfy the environmental norm, output from the treatment unit has to be recycled across it. The wastewater allocation network is shown in Fig. 2b. It may be noted that the entire wastewater at 400 ppm is not passed on to the treatment unit, while some portion of the treated water at 25 ppm is recycled across the treatment unit. Usual rule of designing wastewater allocation network cannot be applied to such problems. However, once the targets are set, nearest neighbor algorithm [38] may be applied to design the wastewater allocation network.

Variation of the minimum effluent treatment flow rate as a function of flow factor for different values of treatment unit outlet concentration is presented in Fig. 3. It may be noted that the constraint related to the minimum allowable inlet concentration to the treatment is relaxed while generating Fig. 3. It may be concluded that for this particular example, flow factor does not play a significant role in increasing the minimum effluent flow rate to be treated in the treatment unit. For this particular example, application of different existing algebraic and graphical methodologies [8], [20], [21], [28] and [31], neglecting flow loss associated with the treatment unit, may not lead to any significant error. However, this may not be the general case and moreover, the total waste discharged to the environment reduces drastically due to higher flow loss (i.e., lower flow factor).

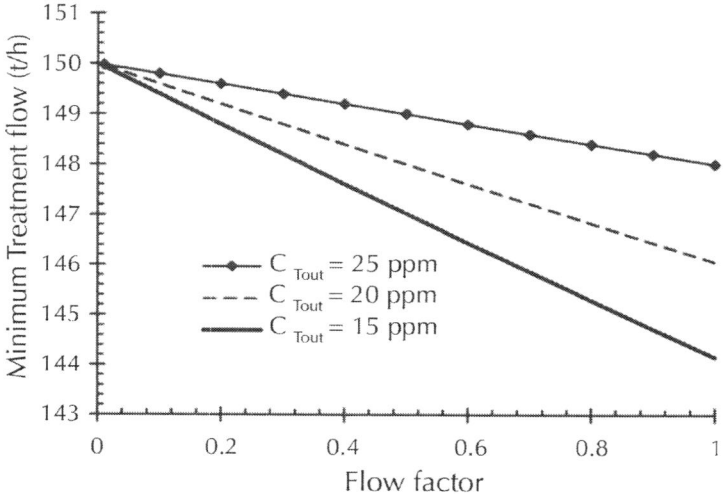

Figure 3: Variation of the minimum effluent treatment flow rate as a function of flow factor for different values of treatment unit outlet concentration.

Example 2: Total Water Management of a Specialty Chemical Plant

This example is also taken from the field of waste management of a specialty chemical plant. Limiting process data for this example are given in Table 4[8] and [20]. The removal ratio and the flow factor of the effluent treatment unit are assumed to be 0.9 and 0.7, respectively. Furthermore, it has been assumed that the treated water cannot be recycled across the water-using processes.

Bandyopadhyay et al. [8] have targeted the minimum requirement of 90.64 t/h of freshwater and 50.64 t/h of wastewater generation. Corresponding freshwater pinch is determined to be 700 ppm. In this example, wastewater is generated at two concentration levels: 20 t/h of wastewater is generated at 1000 ppm and 30.64 t/h of wastewater is generated at 700 ppm.

Table 4: Limiting process data for example 2

Processes	Inlet/demand		Outlet/source	
	Contaminant concentration (ppm)	Flow rate (t/h)	Contaminant concentration (ppm)	Flow rate (t/h)
Reactor/thickener	100	80	1000	20
Cyclone	200	50	700	50
Filtration	0	10	100	40
Steam system	0	10	10	10
Cooling system	10	15	100	5

Concentration of the freshwater, $C_{fw}=0$ ppm. Environmental limit for discharge concentration, C_e = 50 ppm. Characteristics of the treatment unit, α = 0.7 and r = 0.9.

Based on the proposed algorithm (Table 5), the minimum effluent treatment flow rate is targeted to be 129.72 t/h. Source composite curve, waste composite curve, and the treatment line are shown in Fig. 4. Inlet concentration to the treatment unit is calculated to be 350 ppm and the treatment pinch is identified to be 400 ppm. The minimum effluent flow rate to be treated in the treatment unit is more than the total wastewater available. Therefore, to satisfy the environmental norm, output from the treatment unit has to be recycled across the treatment unit. 79.08 t/h of treated water has to be recycled across the treatment unit. Since this is a pinched problem, rules of pinch technology applies and waste allocation network can be designed accordingly (not shown for brevity). Targets for the minimum effluent treatment flow rate, corresponding to the neglected flow loss, is only 86.48 t/h. Application of different existing algebraic and graphical methodologies [8], [20], [21], [28] and [31], neglecting flow loss associated with the treatment unit, underestimates the minimum effluent flow rate to be treated by 33.3%. Treatment line neglecting flow loss is shown in Fig. 4 for visual comparison with the treatment line with flow loss.

Table 5: Waste composite curve and targeting for minimum effluent flow rate for example 2

Contaminant concentration (ppm)	Net flow rate (t/h)	Cumulative flow rate (t/h)	Contaminant mass load (kg/h)	Cumulative mass load (kg/h)	Treatment flow rate (t/h)
1000	20	20	0	0	43.97
700	30.64	50.64	6	6	54.50
50	0	50.64	32.92	38.92	0
0	0	50.64	2.53	41.45	0

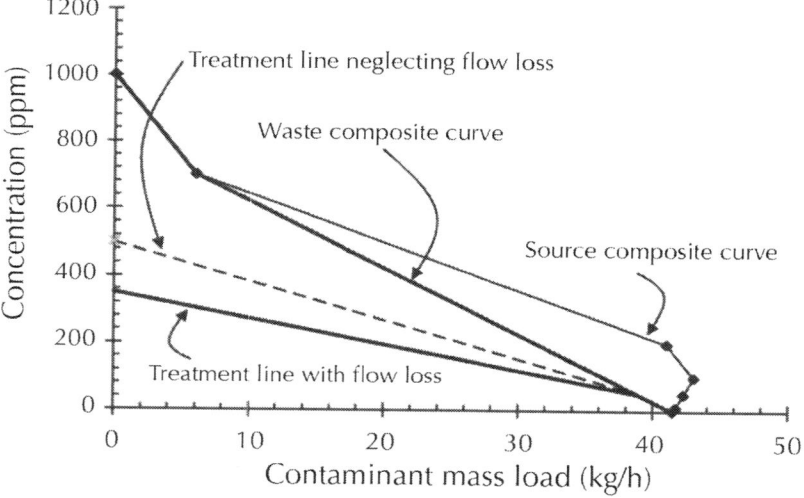

Figure 4: Source composite curve, waste composite curve, and treatment line for example 2.

Variation of the minimum effluent treatment flow rate as a function of flow factor for different values of removal ratio is presented in Fig. 5. It may be concluded that for this particular example, flow factor as well as the removal factor play a significant role in determining the minimum effluent flow rate to be treated in the treatment unit.

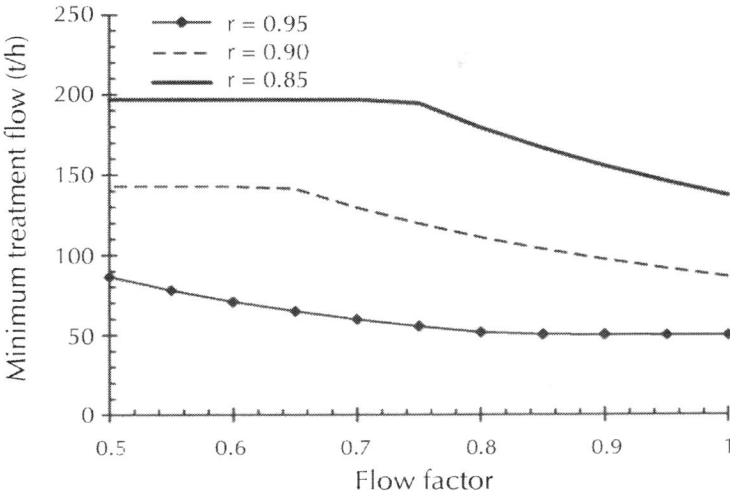

Figure 5: Variation of the minimum effluent treatment flow rate as a function of flow factor for removal factors.

Example 3: Reduction of Emissions of Volatile Organic Compounds

This example is related to the reduction of emissions of volatile organic compounds (VOCs) and VOCs are responsible for producing urban smog. VOCs are emitted from different sources in a process plant: condenser vents, purges, dryers, combustion processes, spillages, tank loading, fugitive emissions from gaskets, shaft seals, sewers, etc. Significant reductions in VOC emissions can usually be achieved by controlling tank venting, condensers and purges and by thorough inspection and maintenance of gaskets and shaft seals. Methods such as condensation, membranes, absorption and adsorption are generally adopted to recover VOC from different source streams. After minimizing VOC emissions from different sources, different recovery processes may be implemented. Recovery, allocation and recycle/reuse of the VOC not only reduce environmental pollution, but also provide significant economic and environmental benefits to the process. Parthasarathy and El-Halwagi [39] presented optimum mass integration strategies for the maximum VOC recovery. After recovery, different treatment units must be considered for reduction in VOC emission. For treatment and reduction in VOCs, incineration in flares, thermal incinerators, catalytic incinerators, biological scrubbers, etc. may be employed. In this example, the minimum VOC laden stream flow rate to be treated in the treatment unit is determined to achieve the environmental norm.

Process data for this example are given in Table 6[40]. Incineration, as the treatment process, removes 99% of the contaminant with a flow factor of 0.75 (assumed). Based on the proposed algorithm (Table 7), the minimum effluent treatment flow rate is targeted to be 9.65 m³/s, which is 4.2% higher due to flow loss. Waste composite curve and the treatment line are shown in Fig. 6. Inlet concentration to the treatment unit is calculated to be 1427.5 mg/m³ and the treatment pinch is identified to be 500 mg/m³.

Table 6: Process data for example 3

#	Concentration (mg/m³)	Flow rate (m³/s)
1	2000	2.5
2	1500	4.5
3	1000	1.4
4	500	1.3
5	200	3.5

Environmental limit for discharge concentration, $C_e = 80$ mg/m³. Characteristics of the treatment unit, $\alpha = 0.75$, and $r = 0.99$.

Table 7: Waste composite curve and targeting for minimum waste treatment flow rate for example 3

Contaminant concentration (mg/m³)	Net flow rate (m³/s)	Cumulative flow rate (m³/s)	Contaminant mass load (mg/s)	Cumulative mass load (mg/s)	Treatment flow rate (m³/s)
2000	2.5	2.5	0	0	6.86
1500	4.5	7	1250	1,250	8.33
1000	1.4	8.4	3500	4,750	9.01
500	1.3	9.7	4200	8,950	9.65
200	3.5	13.2	2910	11,860	9.57
80	0	13.2	1584	13,444	2.27
0	0	13.2	1056	14,500	0

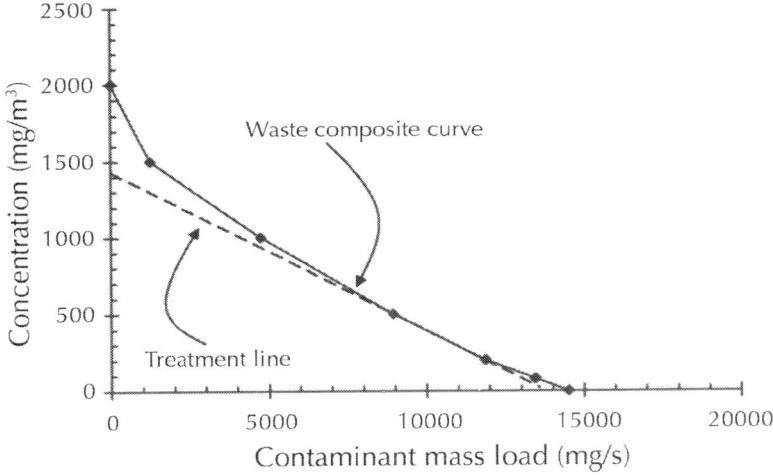

Figure 6: Waste composite curve and treatment line for example 3.

To avoid fire and explosion hazards in vent headers, vent gases are diluted by air or nitrogen. In practice, flammable mixtures are diluted to 30% or less of the flammability limit [40]. By diluting these gases with air or nitrogen, environmental discharge limit, in terms of concentration, can easily be achieved without any treatment unit.

Therefore, it may be more meaningful to set environmental limit as terms of total discharge of waste in the environment. Proposed procedure of targeting cannot be applied directly for such cases. Alternate problem definition and targeting procedure are described below.

TARGETING FOR SPECIFIED TOTAL DISCHARGE

The general problem of targeting the minimum waste treatment flow rate for specified total discharge may be mathematically stated as follows. In a process, a set of N_w waste sources is given. Each waste source produces a flow F_{wi} with a given contaminant concentration of C_{wi}. As the environmental regulations imposed on the overall plant, M_e denotes the total mass flow rate of the contaminant below which waste may be discharged to the environment. Rest of the problem statement is similar to that described in Section2.

The total contaminant load available (M_T) for a given problem can be obtained using Eq. (6). As the environmental discharge limit is known (M_e), total contaminant load removed (M_R) by the treatment unit is expressed as follows:

$$M_R = M_T - M_e \qquad (23)$$

For a treatment unit with constant outlet concentration, Eq. (9) can be used directly to target the minimum waste flow rate to be treated in the treatment unit. In this case, first five steps for generating waste composite curve remain same. However, Eq. (12) in step six should be modified as follows:

$$f_{Tk} = \frac{M_R - M_k}{C_k - \alpha C_{Tout}} \quad \text{when } C_k \geq C_{Tout}$$
$$= 0 \qquad \text{otherwise}$$

$$(24)$$

For a treatment unit with constant removal factor, Eqs. (18) and (19) should be changed suitably to target the minimum waste flow rate to be treated in the treatment unit. For the specified total contaminant flow, the critical contaminant mass load (M_c) is calculated using the following formula:

$$M_c = \frac{M_R(\alpha + r - 1)}{r\alpha}$$

(25)

It may be noted that unlike Eq. (19), the value of the critical mass load is independent of the concentrationCk. To target the minimum flow rate to be treated in the treatment unit with fixed removal ratio and specified total mass load of the contaminant, Eq. (18) should be modified as follows:

$$
\begin{aligned}
f_{Tk} &= \frac{M_R - rM_k}{rC_k} && \text{when } M_k \leq M_c \\
&= \frac{M_R(1-r)/(\alpha r)}{C_{k-1} - (C_{k-1} - C_k)(M_c - M_{k-1})/(M_k - M_{k-1})} && \text{when } M_{k-1} \leq M_c < M_k \\
&= 0 && \text{otherwise}
\end{aligned}
$$

(26)

Applicability of the proposed algorithm is demonstrated through the following flue gas desulphurization example.

Example 4: Flue Gas Desulphurization

Combustion of fuels, containing significant amounts of sulphur, is one of the primary sources of sulphur dioxide (SO_2) emission. SO_2 causes severe damage to the environment and to human health such as urban and industrial decay, acid rain, and pulmonary disease. Environmental regulations are formulated to reduce SO_2 emission. For example, the US Clean Air Act Amendments of 1990 set a goal of reducing annual SO_2 emissions by 10 million tons below 1980 levels. There are many methods available for controlling the emission of SO_2 from boilers. One such method is removal of SO_2 by scrubbing the flue gas with a calcium compound to precipitate calcium sulphate. However, the chloride ion builds up in recirculated scrubbing liquid in the desulphurization unit, should be controlled to reduce corrosion.

Sulphur dioxide (SO_2) emissions from three utility boilers in a chemical process plant are given in Table 8. A wet scrubber-based desulphurization unit with a removal factor of 0.9 and a flow factor of 0.98 has been employed to satisfy the environmental discharge limit of 700×10^6 mg/h. First five steps for generating the waste composite curve are shown in Table 9. Last entry of the fifth column suggests that these three utility boilers produce 5722.5×10^6 mg/h of SO_2. Therefore, applying Eq. (23), it may be determined that 5022.5×10^6 mg/h of SO_2 has to be removed by the desulphurization unit. Results of the targeting step of the proposed algorithm, after replacing Eqs. (18) and (19) with Eqs. (25) and (26), are reported in column six of Table 9. The minimum treatment flow rate of the flue gas is targeted to be 2.221×10^6 N m³/h to satisfy the discharge limit. The treatment pinch point may be identified to be 500 mg/N m³. The waste composite curve and the treatment line are not shown for brevity. It may be noted that conservative performance parameters, related to the removal factor and flow factor, are considered in this example. Removal factors in the range of 0.9–0.98, and flow factors in the range of 0.95–0.99 have been reported based on the actual performance of wet flue gas desulphurization unit [41] and [42]. Finally, SO_2 removed from the flue gas is converted in the form of saleable byproduct, gypsum, a material commonly used in the manufacturing of wallboard. Gypsum is also used, to a lesser extent, as a soil amendment and as an additive in cement [43]. A forced oxidation system is usually employed to oxidize the main reaction product, calcium sulphite, to gypsum.

Table 8: Process data for example 4

#	Concentration (mg/N m³)	Flow rate (N m³/h)
1	5000	900,000
2	4500	105,000
3	500	1,500,000

Environmental limit for total discharge of SO_2, $M_e = 700 \times 10^6$ mg/h. Characteristics of the treatment unit, $a = 0.98$, and $r = 0.90$.

Table 9: Waste composite curve and targeting for minimum flow rate to FGD unit for example 4

Contaminant concentration (mg/N m³)	Net flow rate (10^6 N m³/h)	Cumulative flow rate (10^6 N m³/h)	Contaminant mass load (10^6 mg/h)	Cumulative mass load (10^6 mg/h)	Treatment flow rate (10^6 N m³/h)
5000	0.9	0.9	0	0	1.116
4500	0.105	1.005	450	450	1.140
500	1.5	2.505	4020	4470	2.221
0	0	2.505	751.5	5722.5	2.005

CONCLUSIONS

Recently, environmental concerns have extended to the control of macro- as well as micro- or hazardous pollutants. These pollutants, including globally significant pollutants such as green house gasses, cause significant damage to the ecosystem. Environmental problems span a continuously growing range of pollutants, hazards and ecosystem degradation over wider areas. In this paper, a methodology is proposed to target the minimum waste flow to the treatment unit to satisfy the environmental discharge condition. In the proposed methodology, the conventional models of a treatment unit, with constant outlet concentration and fixed removal ratio, are extended with flow loss that is proportional to the inlet flow rate. Applicability of the proposed methodology is demonstrated through illustrative examples from the domain of water management, volatile organic compound treatment and flue gas desulphurization. The environmental safe limits for discharge for various pollutants are defined either as concentration or total mass flow rate. Based on the ecosystem degradation potential of various pollutants, regulation norms for safe discharge may be defined and accordingly the proposed methodology may be generalized.

The proposed methodology is based on an algebraic approach, calculated on a tableau. An associated graphical representation is also provided to gain physical insight of the targeting problem. However, the proposed methodology is restricted to single contaminant only. Current research is directed towards developing appropriate targeting technology for multiple contaminants. The proposed methodology is also restricted for a single treatment unit. Due to stringent environmental regulations, it may not always be possible to satisfy them with a single treatment unit. Multiple treatment units may have to be employed to satisfy such stringent environmental safe discharge limits. Current research is also directed towards developing appropriate targeting technology for optimal networks of multiple treatment units.

REFERENCES

1.   M.M. El-Halwagi, Process Integration, Academic Press, Elsevier, 2006.

2.   I.C. Kemp, Pinch Analysis and Process Integration: A User Guide on Process Integration for the Efficient Use of Energy, Elsevier, 2007.

3.   S. Bandyopadhyay, M. Mishra, U.V. Shenoy, Energy-based targets for multiplefeed distillation columns, AIChE J. 50 (2004) 1837–1853.

4.   J. Varghese, S. Bandyopadhyay, Targeting for energy integration of multiple fired heaters, Ind. Eng. Chem. Res. 46 (2007) 5631–5644.

5.   M.M. El-Halwagi, V. Manousiouthakis, Synthesis of mass exchange networks, AIChE J. 8 (1989) 1233–1244.

6.   N. Hallale, D.M. Fraser, Capital cost targets for mass exchange networks. A special case: water minimization, Chem. Eng. Sci. 53 (1998) 293–313.

7.   Y.P. Wang, R. Smith, Wastewater minimization, Chem. Eng. Sci. 49 (1994) 981–1006.

8.   S. Bandyopadhyay, M.D. Ghanekar, H.K. Pillai, Process water management, Ind. Eng. Chem. Res. 45 (2006) 5287–5297.

9.   H.K. Pillai, S. Bandyopadhyay, A rigorous targeting algorithm for resource allocation networks, Chem. Eng. Sci. 62 (2007) 6212–6221.

10. G.N. Kulkarni, S.B. Kedare, S. Bandyopadhyay, Determination of design space and optimization of solar water heating systems, Solar Energy 81 (2007) 958–968.

11. A. Roy, P. Arun, S. Bandyopadhyay, Design and optimization of renewable energy based isolated power systems, SESI J. 17 (2007) 54–69.

12. P. Arun, R. Banerjee, S. Bandyopadhyay, Sizing curve for design of isolated power systems, Energy Sustain. Dev. 11 (2007) 21–28.

13. G.N. Kulkarni, S.B. Kedare, S. Bandyopadhyay, Design of solar thermal systems utilizing pressurized hot water storage for industrial applications, Solar Energy 82 (2008) 686–699.

14. G.N. Kulkarni, S.B. Kedare, S. Bandyopadhyay, Optimization of solar water heating systems through water replenishment, Energy Convers. Manage. 50 (2009) 837–846.

15. R.R. Tan, D.C.Y. Foo, Pinch analysis approach to carbon-constrained energy sector planning, Energy 32 (2007) 1422–1429.

16. S.C. Lee, D.K.S. Ng, D.C.Y. Foo, R.R. Tan, Extended pinch targeting techniques for carbon-constrained energy sector planning, Appl. Energy 86 (2009) 60–67.

17. P. Arun, R. Banerjee, S. Bandyopadhyay, Optimum sizing of battery integrated diesel generator for remote electrification through design-space approach, Energy 33 (2008) 1155–1168.

18. S. Bandyopadhyay, Source composite curve for waste reduction, Chem. Eng. J. 125 (2006) 99–110.

19. N. Takama, T. Kuriyama, K. Shiroko, T. Umeda, Optimal water allocation in petroleum refinery, Comput. Chem. Eng. 4 (1980) 251.

20. Y.P. Wang, R. Smith, Design of distributed effluent treatment systems, Chem. Eng. Sci. 49 (1994) 3127.

21. W.C.J. Kuo, R. Smith, Effluent treatment system design, Chem. Eng. Sci. 52 (1997) 4273.

22. B. Galan, I.E. Grossmann, Optimal design of distributed wastewater treatment networks, Ind. Eng. Chem. Res. 37 (1998) 4036.

23. R. Hernández-Suaˇırez, J. Castellanos-Fernández, J.M. Zamora, Superstructure decomposition and parametric optimization approach for the synthesis of distributed wastewater treatment networks, Ind. Eng. Chem. Res. 43 (2004) 2175.

24. I.S.F. Freitas, R.A.R. Boaventura, C.A.V. Costa, Conceptual design of industrial wastewater treatment processes: a hierarchical approach procedure, in: Proceedings of the Second Conference on Process Integration, Modeling and Optimization for Energy savings and Pollution Reduction (PRESS'99), Budapest, Hungary, 1999.

25. J.M. Douglas, Conceptual Design of Chemical Processes, McGraw-Hill, New York, 1988.

26. G. Statyukha, O. Kvitka, I. Dzhygyrey, J. Jezowski, A simple sequential approach ˙ for designing industrial wastewater treatment networks, J. Clean. Prod. 16 (2008) 215–224.

27. T.K. Zhelev, N. Bhaw, Combined water–oxygen pinch analysis for better wastewater treatment management, Waste Manage. 20 (2000) 665–670.

28. W.C.J. Kuo, R. Smith, Designing for the interactions between water use and effluent treatment, Trans. Inst. Chem. Eng. 76 (1998) 287.

29. C.-T. Chang, B.-H. Li, Improved optimization strategies for generating practical water-usage and -treatment network structures, Ind. Eng. Chem. Res. 44 (2005) 3607–3618.

30. M. Gunaratnam, A. Alva-Argaˇı ez, A. Kokossis, J.-K. Kim, R. Smith, Automated design of total water systems, Ind. Eng. Chem. Res. 44 (2005) 588–599.

31. D.K.S. Ng, D.C.Y. Foo, R.R. Tan, Targeting for Total Water Network. Part 2. Waste treatment targeting and interactions with water system elements, Ind. Eng. Chem. Res. 46 (2007) 9114–9125.

32. S. Bandyopadhyay, C.C. Cormos, Water management in process industries incorporating regeneration and recycle through single treatment unit, Ind. Eng. Chem. Res. 47 (2008) 1111–1119.

33. D.K.S. Ng, D.C.Y. Foo, R.R. Tan, Y.L. Tan, Ultimate flowrate targeting with regeneration placement, Chem. Eng. Res. Design 85 (9) (2007) 1253–1267.

34. A. Alva-Argáez, A.C. Kokossis, R. Smith, Automated design of industrial water networks. American Institute of Chemical Engineering Annual Meeting, Paper 13f, Miami, FL, 1998.

35. A. Alva-Argáez, A.C. Kokossis, R. Smith, Wastewater minimisation of industrial systems using an integrated approach, Comput. Chem. Eng. 22 (1998) S741–S744.

36. R.R. Tan, D.K.S. Ng, D.C.Y. Foo, K.B. Aviso, A superstructure model for the synthesis of single-contaminant water networks with partitioning regenerators Process Saf. Environ. Protect. 87 (2009) 197–205.

37. D.K.S. Ng, D.C.Y. Foo, R.R. Tan, C.H. Pau, Y.L. Tan, Automated targeting for conventional and bilateral property-based resource conservation network, Chem. Eng. J. 149 (2009) 87–101.

38. R. Prakash, U.V. Shenoy, Targeting and design of water networks for fixed flowrate and fixed contaminant load operations, Chem. Eng. Sci. 60 (2005) 255–268.

39. G. Parthasarathy, M.M. El-Halwagi, Optimum mass integration strategies for condensation and allocation of multicomponent VOCs, Chem. Eng. Sci. 55 (2000) 881–895.

40. C.-W. Hui, R. Smith, Targeting and design for minimum treatment flowrate for vent streams, Trans. Inst. Chem. Eng. 79 (2001) 13–24.

41. C. Erickson, M. Jasinski, L. VanGansbeke, Wet flue gas desulfurization (WFGD) upgrade at the trimble county generating station unit 1, in: Proceedings of the MEGA Symposium, Baltimore, Maryland, August 28–31, 2006.

42. M. Quitadamo, C. Erickson, J. Langone, SO2 removal enhancement to the vectren, culley generating station units 2&3 wet flue gas desulfurization system, in: Proceedings of the ICAC Forum 05', Baltimore, Maryland, March 7–10, 2005.

43. C.L. Kairies, K.T. Schroeder, C.R. Cardone, Mercury in gypsum produced from flue gas desulfurization, Fuel 85 (17–18) (2006) 2530–2536.

5

Intensified Processes for FAME Production from Waste Cooking Oil: A Technological Review

Alex Mazubert, Martine Poux, and
Joëlle Aubin

Université de Toulouse, INP, LGC (Laboratoire de Génie Chimique),
4 Allée Emile Monso, BP-84234, F-31030 Toulouse, France CNRS,
LGC, F-31030 Toulouse, France

ABSTRACT

This article reviews the intensification of fatty acid methyl esters (FAME) production from waste cooking oil (WCO) using innovative process equipment. In particular, it addresses the intensification of WCO feedstock transformation by transesterification, esterification and hydrolysis reactions. It also discusses catalyst choice and product separation. FAME production can be intensified via the use of a number of process equipment types, including as cavitational reactors, oscillatory baffled reactors, microwave reactors, reactive distillation, static mixers and microstructured reactors. Furthermore, continuous flow equipment that integrates both reaction and separation steps appear to be the best means for intensifying FAME production. Heterogeneous catalysts have also shown to provide attractive results in terms of reaction performance in certain equipment, such as microwave reactors and reactive distillation.

INTRODUCTION

Context and Objectives

Biological resources are currently the best alternative to fossil fuels or petrochemical solvents for renewable energy and green chemistry applications. The production of biodiesel, composed of fatty acid methyl esters (FAME), has received much attention in research over the last 15 years as it is an organic, biodegradable and non-toxic fuel source that is made from renewable resources such as vegetable oils and animal fats. Although virgin and food-grade oils have proven to be suitable feedstocks for biodiesel production, the use of edible vegetable oils evokes the 'food versus fuel' debate on the use of widespread farmland areas for biofuel production in detriment of food supply [1]. Furthermore, in addition to increasing food demands the increasing demand for oil for biofuel production could ultimately lead to deforestation and desertification [2]. The use of non-edible crops, such as jatropha or castor oil, avoids the direct competition for food oils but does not resolve the problem of requiring large plantation land areas [3]. This competition explains the high price of edible oils, representing about 60–90% of the process cost [4]. Waste cooking oil (WCO) as a potential renewable feedstock appears to be an economically and environmentally viable solution for FAME production and presents a number of advantages: e.g. WCO are two to three times cheaper than virgin oils [5], the recycling of WCO reduces waste treatment costs [6], and the quality of the FAME is the same as that produced with virgin oils [7]. In France, the release of waste oils in sewage is prohibited and governmental legislations require waste cooking oils from restaurants and food industries to be collected for recycling and disposal (France, article R. 1331-2, public health code [8]). In the UK, the collect of waste cooking oil is strongly supported for reducing costs of waste disposal and for the production of renewable energy (Food standards Agency [9]). In the USA, several companies and communities (e.g. Restaurant Technologies Inc. [10], Shakopee Mdewakanton Dakota Community [11]) provide solutions for waste oil management and recycling.

Although the main use of FAME is biodiesel, many other industrial applications exist [12]. Indeed, FAME possesses good solvent properties with low volatility, as well as being biodegradable and non-toxic. As a result, FAME have been used to wash metal pieces [13], printing material, graffiti [14], automobiles and plane parts [15], as well as for cleaning up oil spills [16]. FAME is also used as binders in inks [17], as well as thinning agents for building and civil engineering work [18]. Other applications include the production of pesticides [19] and phytosanitary products [20].

Numerous publications on FAME production are available in the current literature and most of these studies focus on various aspects of the transesterification reaction. Amongst these, a number of reviews have concentrated on the different ways to catalyze the transesterification reaction [21], [22], [23], [24], [25] and [26] and particularly with calcium oxide [27] or heterogeneous catalysis [28], [29], [30] and [31]. Some papers have also focused on the different means for pre-treating WCO before transesterification [6], [7], [21] and [23]. Qiu et al. [32] reviewed process intensification technologies for biodiesel production via homogenous base-catalyzed transesterification. Talebian-Kiakalaieh et al. [33] and Maddikeri et al. [34] reviewed novel biodiesel processes using WCO feedstocks. Emerging processes for biodiesel production are also reviewed by Oh et al. [35]. A number of authors also discuss acoustic cavitation, microwaves [22], [24], [26] and [36], membrane reactors and reactive distillation [37] as means to intensify the transesterification reaction for FAME production. Finally, the methods for separation and purification of biodiesel at the outlet of the reactor have been discussed by Leung et al. [21], Enweremadu and Mbarawa[23], and Atadashi et al. [38].

The objective of this article is to critically review the different intensified process equipment available and adapted to the global FAME production process and to evaluate their use for the transesterification reaction, as well as the pretreatment reactions (i.e. esterification and hydrolysis) in the case where WCO is used as feedstock. In particular, reaction performance, mass transfer enhancement, ease of separation, possible catalyst types (acid or base, homogeneous or heterogeneous) and energy efficiency of the equipment are discussed. Finally, recommendations on the choice of intensified process equipment for FAME production are given.

In this review the discussion and evaluation of the capacities of various intensified process equipment for FAME production (Sections 2 and 3) is preceded with technical information on the related reactions, catalysis, WCO regeneration, and conventional industrial processes (Sections 1.2, 1.3 and 1.4) as introductory material.

Reactions and Catalysis

In the biodiesel industry, vegetable oil is usually transformed into FAME by transesterification. In this reaction, triglyceride reacts with alcohol to give FAME and glycerol, as shown in Fig. 1. Typically, methanol is preferred as the alcohol because of its higher reactivity and low price [21].

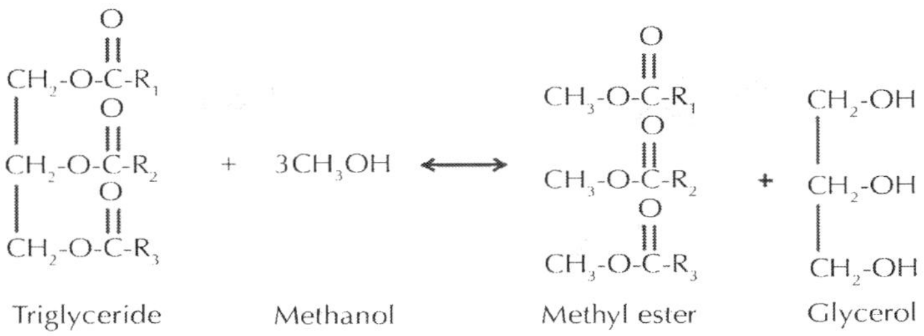

Figure 1: Transesterification reaction scheme.

This reaction is catalyzed using either homogeneous or heterogeneous, acid or basic catalysts, or enzymes. Various catalysts exist and have been well described in a number of recent reviews [21], [22], [23], [24], [25], [26] and [36]. Table 1 summarizes the strengths and weaknesses of the different catalyst types. In industrial FAME processes, homogeneous alkali catalysts (e.g. KOH, NaOH) are the most often used. These inexpensive catalysts lead to short reaction times and are easy to handle in terms of transportation and storage [36]. However, base-catalyzed transesterification is very sensitive to the presence of free fatty acids (FFA) in the oil, which leads to undesired soap formation. Soap formation decreases the reaction yield and facilitates the formation of emulsions, thereby causing difficulties in the downstream separation process [3] and [39]. Consequently, it is generally recommended that the acidity of the oil must be less than 1 mg KOH/g oil [6] and [23]. Alkali catalysis is also sensitive to presence of water, forming inactive alkaline soaps [40]. Water content is therefore limited to 0.05 vol. % (ASTM D6751 standard) [21]. It is important to point out that the FFA content is very different in unused vegetable oils compared with WCO. Cooking processes and the presence of water cause the hydrolysis of triglycerides, which increases the FFA content in the oil [3] and [41]. The absence of oxygen at high temperatures causes thermolytic reactions whereas oxidative reactions occur if air is dissolved [22], thereby changing the composition of the oil. Polymerization and saponification can also occur [7]. Furthermore, the viscosity, surface tension and color of virgin oils change after the cooking processes [42]. Unlike base-catalyzed transesterification, acid-catalyzed transesterification is insensitive to FFA content. However, it is associated with a number of disadvantages, for example the need of higher reaction temperatures and alcohol to oil molar ratios, the difficulty of catalyst separation, as well as serious environmental and corrosion issues [22]. Most importantly, the reaction rate of acid-catalysed transesterification has been reported to be 4000 times slower than those using alkalis [43]. This has been proven in a number of studies: e.g. a base-catalyzed transesterification carried out at 65 °C, a methanol to oil molar ratio of 6:1, and 1% w/w of KOH has shown to give a 96.15% yield in 1 h [44] whereas an acid-catalyzed transesterification carried out at 65 °C, a methanol to oil molar ratio of 30:1% and 1% w/w of sulfuric acid has shown to give 90% yield in 69 h [45].

Table 1: Comparison of the different types of catalysis of the transesterification reaction [22]

Type of catalyst	Advantages	Disadvantages
Homogenous base	– Very fast reaction rate	– Sensitive to FFA content
	– Mild reaction conditions	– Soap formation (causing yield to decrease and increase difficulty for product and catalyst separation)
	– Inexpensive	
Heterogeneous base	– Faster than acid– catalyzed reaction	– Poisoned at ambient air
	– Mild reaction conditions	– Sensitive to FFA content
	– Easy separation of catalyst	– Soap formation
	– Easy reuse and regeneration of catalyst	– Leaching of catalyst causing contamination of product
		– Energy intensive
Homogeneous acid	– Insensitive to FFA and water content	– Very slow reaction rate
	– Simultaneous esterification and transesterification possible	– Corrosive catalysts (e.g. H_2SO_4)

	– Mild reaction conditions	– Separation of catalyst is difficult
Heterogeneous acid	– Insensitive to FFA and water content	– Complicated reaction synthesis leading to higher processing costs
	– Simultaneous esterification and transesterification	– High reaction temperature, high alcohol to oil molar ratio, long reaction times
	– Easy separation of catalyst	– Energy intensive
	– Easy reuse and regeneration of catalyst	– Leaching of catalyst causing contamination of product
Enzyme	– Low reaction temperature (lower than for homogenous base catalysts)	– Very slow reaction rate
	– Only one purification step necessary	– High costs
		– Sensitive to alcohol (typically methanol, causing deactivation)

As summarized in Table 1, heterogeneous catalysis offers a number of advantages over homogenous catalysis and in particular, the ease of catalyst separation and the possibility to reuse and regenerate it. However, in industrial practice, heterogeneous catalysts are less widely used due to the high temperatures required, the problems related to catalyst leaching, and their sensitivity to FFA and water content [46]. Nevertheless, recently studies show that a novel heterogeneous catalyst, srontium oxide (SrO), gives faster reaction rates than the usual homogenous catalyst types, such as potassium hydroxide [46] and [47]. The yield is greater than 95% at 65 °C in 30 min of reaction time in presence of 3 wt.% of SrO and the activity is retained for 10 cycles [47]. A SrO/SiO_2 catalyst leads to a 95% conversion in 10 min and even with about 3 wt. % of FFA and water, the conversion is still greater than 90% in 20 min [46]. One of the main advantages of this novel alkali heterogeneous catalyst is that it allows WCO treatment with less strict requirements: indeed, SrO can catalyze the transesterification of WCO even if the FFA and water content are greater than the limits required for conventional homogeneous base-catalysts (3 wt.%).

Regeneration of Waste Cooking Oil

Since the interest in WCO for biofuel production has increased in recent years, there is an increased need for the reduction of FFA in WCO. Indeed, if the FFA content of WCO is too high (superior to 1 mg KOH/g oil), the transesterification conversion rate is insufficient and too much soap is produced, hindering the separation of glycerol and ester. In such cases, a pretreatment step is required to regenerate the WCO, thereby improving its quality in terms of FFA and water content in detriment of increased processing costs [23]. As described in a recent review [36], many physical methods for reducing FFA, moisture and solids content in WCO exist. Moisture can be removed by drying (with $MgSO_4$, calcium chloride, ion exchange resins), by filtration (under vacuum, through a chromatography column, silica gel or cellulose fiber, under microwave irradiation), with steam injection and sedimentation treatment, as well as by distillation. FFA content can be reduced by the neutralization and the separation of soaps in a decanter, membrane filtration [48], and solvent extraction (particularly with anhydrous methanol or ethanol) [49]. Filtration is typically used for the removal of suspended solids and water washing is employed for the separation of soluble salts from the WCO [6].

Chemical pretreatment methods are another way to reduce FFA content. Three alternatives exist for FAME production. The first is acid-catalyzed esterification shown in Fig. 2. The esterification of FFA in WCO is a solution to reduce the level of FFA in the oil before performing the transesterification.

$$HO\text{-}\overset{\overset{\displaystyle O}{\|}}{C}\text{-}R \quad + \quad CH_3OH \quad \longleftrightarrow \quad CH_3\text{-}O\text{-}\overset{\overset{\displaystyle O}{}}{C}\text{-}R \quad + \quad H_2O$$

| Fatty acid | Methanol | Methyl ester | Water |

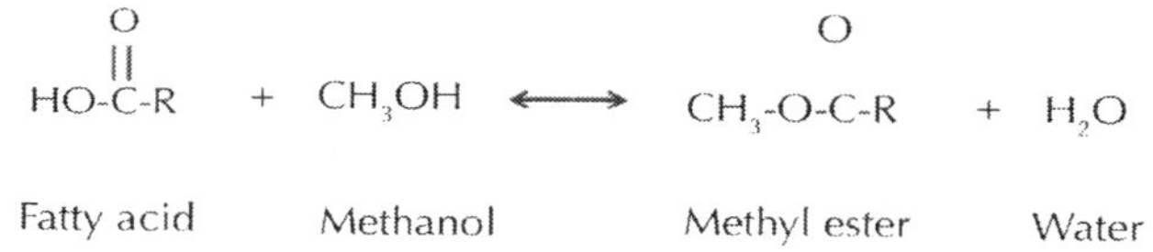

Figure 2: Reaction scheme for the esterification of fatty acids with methanol to methyl esters and water.

Esterification typically uses methanol (or ethanol) but glycerol may also be employed [50] as shown by the reaction scheme, given in Fig. 3. This direct esterification of fatty acids with glycerol is of particular interest since the WCO is not only regenerated but the FFA levels are also reduced using glycerol, which is the side-product of the biodiesel reaction. Another industrial application of this reaction is production of monoglycerides [51]. A number of recent studies have focused on the development of various heterogeneous catalysts for this reaction including zinc [50], Fe–Zn double-metal cyanide catalyst [52], and solid superacid SO^{2-}_4 / ZrO_2 – Al_2O_3 [53] leading to acceptable conversion at 200 °C and under atmospheric pressure after 3–4 h reaction time. Enzymatic catalysis has also been explored and has shown to give about 90% conversion in about 80 h with immobilized lipase (Rhizomucor miehei) [54] and a 96.5% yield is obtained in 72 h with Novozym 435 [55].

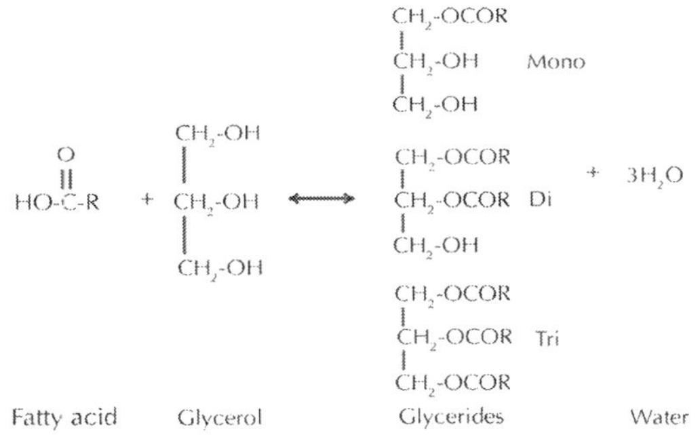

Figure 3: Reaction scheme of the esterification of fatty acids with glycerol to glycerides and water.

Another chemical route for the regeneration of WCO and the production of FAME is hydroesterification. In this reaction the, FFA are first concentrated via hydrolysis as shown in Fig. 4, and then transformed to FAME via esterification. Continuous hydrolysis processes have existed in industry for many years (e.g. Colgate-Emery, Foster-Wheeler). They are typically conducted at high temperature (260 °C) due to the absence of catalyst and at high pressures (50 bars) to maintain the reactants in a liquid state, enabling thus 98% conversion [56] and [57]. More recently, Satyarthi et al. [58] obtained 80% of FFA at 190 °C with a solid catalyst (Fe–Zn) in 12 h. Hydroesterification thus combines the well-known hydrolysis process with fatty acid esterification. It is currently used in commercial biodiesel plants in Brazil [59].

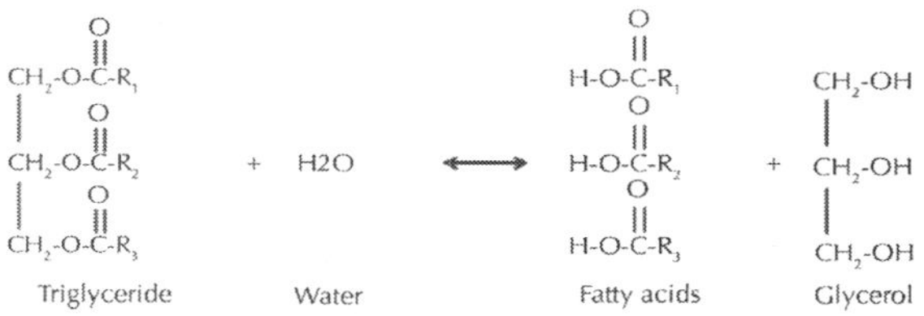

Figure 4: Reaction scheme of the hydrolysis of triglycerides to fatty acids and glycerol.

It is interesting to point out that all three reactions related to FAME production (i.e. transesterification, esterification and hydrolysis) involve a reaction between two immiscible reactants and are mass-transfer limited. Further, a numbers of similarities can be found between these three reaction types. A comparison of performance of these different reactions in conventional processes is given in the next section.

Conventional Processes

Before entering the discussion on the various intensified process equipment for FAME production, it is important to highlight the characteristics of existing FAME production or related processes using conventional equipment, such as stirred tank reactors, packed beds and columns. Table 2 gives typical examples of process characteristics of transesterification and esterification reactions performed in laboratory-scale stirred tank reactors. The most important features of these reactions are the immiscibility (or partial miscibility) of the reactants and products, and the reaction time for acceptable conversion, which is of the order of a couple of hours in these small scale reactors.

Table 2: Comparison of the characteristics of the different reactions in conventional processes [50], [60] and [61]

Reaction	Catalyst	Reactor	Miscibility			Operation conditions			Reaction time and conversion		Reference
			Reactants	During the reaction	Products	T(°C)	P(atm)		t(h)	c(%)	
Transesterification	Base/Homogenous	1.5 L glass cylindrical reactor	No	Yes	No	60	1		1	88	[60]
Esterification (with Methanol)	Acid/Homogenous	0.5 L three neck round bottom flask	Low	Low	Low	60	1		1.5	96.6	[61]
Esterification (with Glycerol)	Acid/Heterogeneous	Lab scale stirred reactor	No	Partial (increased with high temperatures)	No	200	1		3	90	[50]

At the level of industrial production, the French Institute of Petroleum (IFP) has developed both a discontinuous and a continuous process for transesterification using a homogenous alkali catalyst [62]. The discontinuous reactor operates at temperatures between 45 and 85 °C with a maximum pressure of 2.5 bars and produces 80,000 tons of FAME per year. The time required to reach thermodynamic equilibrium is on the scale of 1 h and the yield is between 98.5% and 99.4%. The continuous process leads to similar yields and the production capacity is superior to 100,000 tons per year. A continuous heterogeneous base-catalyzed transesterification process named ESTERFIP-H™ has also been developed by IFP [62]. The catalyst is a zinc aluminate spinel (Zn Al$_2$ O$_4$), the operating temperature and pressure ranges between 180–220 °C and 40–60 bar, respectively. The yields achieved are greater than 98%. In Brazil, there are a significant number of biodiesel plants in operation. One example is the biodiesel plant in Belém-PA-Brazil that started up in April 2005. This plant produces 12,000 ton/year using a heterogeneous catalytic esterification of fatty acids in a fixed bed reactor and involves several steps of chemical reaction and product separation [59].

Examination of the literature data on the performance of laboratory and industrial processes highlights that transesterification, esterification and hydrolysis are all limited by mass transfer. In the case of transesterification, the mass transfer limitation [63] results in reaction times of the order of a couple of hours for desired conversion (up to 96.5%) [60] and [64]. Indeed, the reaction rate is limited by the immiscibility of triglycerides and methanol at the beginning of the reaction and then because the glycerol phase separates out taking most of the catalyst with it as the reaction proceeds [66]. This latter point is confirmed by kinetics, which shows that the reaction rate is sigmoidal, i.e. slow at both the beginning and end of the reaction but fast at an intermediate stage [67]. Indeed, the initial immiscibility between reactants rapidly disappears because of the formation of diglycerides and monoglycerides, which play the role of emulsifier [68]. As the mono and diglycerides appear, the size of the dispersed phase droplets decreases and then increases as glycerol is formed [67]. The size of droplets (and consequently the interfacial area) therefore directly influences the reaction rate [69]. Esterification with both methanol and glycerol is limited by mass transfer due to the low solubility between the reactants [70] and [71]. Partial miscibility is observed with short-chain fatty acids and equilibrium can be shifted with an excess of reactant or the removal of the product. In hydrolysis, high temperatures allow higher oil solubility in water and a better electrolytic dissociation of water, both of which improve mass transfer. Equilibrium can also be shifted using an excess in water [56] or by removing product.

The use of process intensification equipment for FAME production provides a means to reduce the mass transfer limitations related to transesterification, esterification and hydrolysis with less energy consumption, as well as under safer and cleaner conditions. A step towards integrating process intensification in biofuel plants is illustrated by the Biobrax plant (Bahia, Brazil), which produces 60,000 tons of biodiesel per year by hydroesterification [72]. The process is based on a combination of a counter current splitting column and reactive distillation with a heterogeneous catalyst operated at 260 °C. Although distillation is an energy demanding operation, the reactive distillation step has the advantage of combining the reaction and separation steps of the esterification, providing thus a more efficient and cleaner process. Indeed, it has been shown under batch conditions, where the reaction and separation steps are carried out consecutively, that the time required to separate the ester product from the glycerol is sixteen times greater than the time required reach 95% conversion [44].

PROCESS EQUIPMENT FOR THE INTENSIFICATION OF FAME PRODUCTION

A number of means to intensify processes exist and these involve a range of equipment types and methods, including microreaction technology, multifunctional reactors and novel activation techniques like microwaves and ultrasounds [73] and [74]. In this section, the use of process intensification techniques that are adapted to liquid–liquid mass transfer limited reactions, such as transesterification, esterification and hydrolysis are reviewed and assessed. In the following sections, yield is defined by the relation:

$$y = n_{FAME}/3n_{TG}$$

$$(1)$$

y is the yield, n_{FAME} is the number of moles of FAME and n_{TG} is the initial number of moles of triglycerides.

Conversion is defined by the relation:

$$c = (n_{TG\ initial} - n_{TG\ final})/n_{TG\ initial}$$

$$(2)$$

c is the conversion (%), $n_{TG\ initial}$ is the initial number of moles of triglycerides and $n_{TG\ final}$ is the final number of moles of triglycerides.

Microreactors

Microreaction technology for chemical and biological applications has undergone major technical and scientific development since the mid-1990s and these miniaturized reaction systems have now proven, through both research and industrial practice, to provide innovative and sustainable solutions for the chemical and process industries [75] and [76]. Due to the small characteristic dimensions and the extremely high surface to volume ratio of these reactors, heat and mass transfer are remarkably intensified and the temperature within can be tightly controlled. These features make microreactors particularly adapted to mixing limited, highly exo/endothermic and/or mass transfer limited reactions.

Microreactors are of particular interest for performing immiscible liquid–liquid reactions because they potentially offer very high interfacial area between phases, thereby improving the rate of mass transfer [77]. Indeed, mass transfer enhancement is highly dependent on the liquid–liquid flow regime and droplet size, which depend not only on the physical properties of the fluids (viscosity, density, interfacial tension…) but also the operating conditions, the reactor geometry, as well as the properties of the construction material (e.g. wettability, roughness) [76]. In liquid–liquid flow, two types of flow regimes typically occur, namely slug-flow and parallel flow as depicted in Fig. 5[77] and [78]. The slug-flow pattern is characterized by drops of uniform size, separated by uniform lengths of the continuous phase. Two mechanisms contribute to the mass transfer process: convection within the dispersed drops and the continuous phase slugs due to the creation of recirculation flow patterns [77] and [79] and diffusion between adjacent slugs of dispersed and continuous phases. Mass transfer can be enhanced by increasing the interfacial area and the intensity of recirculation in the slugs, which depend on operating conditions. Parallel flow consists of two continuous streams of the dispersed and continuous phases flowing parallel to one another, without convective mixing. Here, the rate of mass transfer is governed by the diffusive mechanism and depends directly on the characteristic size of the fluid streams and therefore the microreactor geometry. The fact that no droplets are formed means that phase separation at the reactor outlet is relatively easy [80]. It is worthwhile point out that the interfacial area and intensity of internal recirculation in slug flow can be modified by changing the flow rate in a given microreactor, whereas in parallel flow the interfacial area is determined by the characteristic dimension of the microreactor geometry or microchannel.

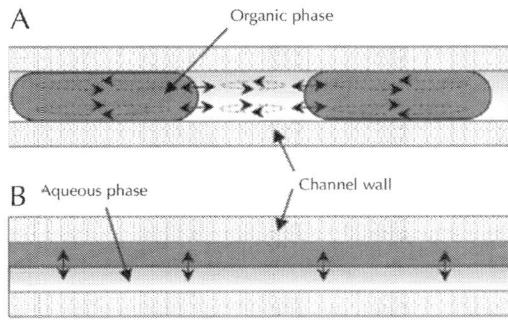

Figure 5: Types of liquid–liquid flow patterns in microreactors: (A) slug flow; and (B) parallel flow [78].

For the transesterification reaction, a microreactor may be a viable choice because the reaction is limited by mass transfer and the small characteristic size of microreactors allows high interfacial areas between phases to be obtained. In transesterification, the reactants (e.g. methanol and glycerides) are immiscible at the beginning of the reaction, an emulsion of fine droplets (or pseudo-homogenous phase) is obtained during the reaction, and then at the end of the reaction the products (esters and glycerol) form two distinct and immiscible phases again [69]. The emulsion of fine droplets formed during the course of the reaction (depicted in Fig. 6) is specific to transesterification and due to the formation of mono and diglycerides [68].Table 3 shows a typical advantage of microreactors over batch reactors in the fact that a smaller emulsion droplet size is achieved in a shorter processing time [81], which means that higher conversion in a shorter time can be achieved in microreactors.

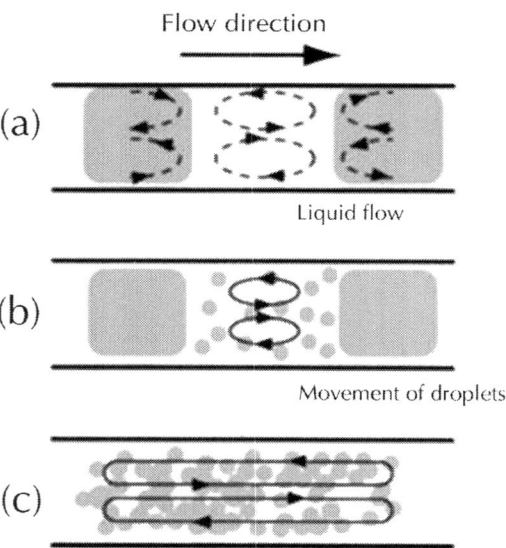

Figure 6: Evolution of liquid–liquid flow patterns over the course of a transesterification reaction in a microreactor. (a) Slug flow whereby the reactants form two distinct phases; (b) Beginning of emulsion formation

in the continuous phase due to the production of mono- and di-glycerides; and (c) Emulsion during the course of the reaction, before phase separation of products. [82].

Table 3: Influence of the type of reactor on the droplets mean diameter (dm) for the transesterification reaction [94] and [81]

Type of reactor	Operating time (min)	dm (μm)
Batch (250 mL)	3	12.9
(two flat-blade paddle agitator)	60	5.1
Microreactor (d = 140 μm)	0.5	1.9

A number of studies in the literature have demonstrated the feasibility of using microreactors for improving the performance of transesterification reactions. Table 4 compares the different data available for a homogeneous base-catalysed transesterification reaction in terms of reaction and flow conditions, microreactor characteristics and reaction performance.

Table 4: Results obtained for homogeneous base-catalyzed transesterification in microreactors. Y: yield (%), c: conversion (%), R: molar ratio methanol:oil, t: residence time (min), w: weight fraction of catalyst (%), T: temperature (°C), : reactor dimension (mm), F: flow-rate (mL/h). I: beginning of the reaction, II: during the reaction, III: end of the reaction, s-f: slug-flow, e: emulsion, p: parallel flow [81], [82], [83], [84] and [85]

Oil	Catalyst	Performance		Reaction conditions				Type of reactor	Flow-pattern			Material	φ	F	Reference
		y	c	R	t	w	T		I	II	III				
Cottonseed (0.8% FFA)	KOH	99.4	–	6	6	1	60	Micro-mixer + Tube	s-f	–	s-f	Quartz	0.25	14.7	[83]
Soybean	NaOH	99.5	–	9	0.47	1.2	56	Zigzag micro-channel	e	e	e	Stainless steel	0.24	8.1	[81]
Canola	NaOH	–	99.8	6	3	1	60	T-mixer + Tube	–	s-f	p	Teflon	1.5	231	[84]
Sunflower	KOH	–	100	23.9	1	4.5	60	T-joint + tube	s-f	e	e	FEP	0.8	8.2	[82]
		–	59.3	4.6	3.7	4.5	60	T-joint + tube	s-f	e	s-f	FEP	0.8	8.2	[82]
		–	100	23.9	1.6	4.5	40	T-joint + tube	s-f	e	s-f	FEP	0.8	8.2	[82]
		–	97	11.3	0.83	4.5	60	T-joint + tube	–	–	–	FEP	0.8	8.2	[82]
Soybean	KOH	–	100	6	1.5	3.32	60	T-joint + Slit-channel	s-f	s-f	s-f	Nylon	2 × 152.4 × 1	12.2	[85]

From the ensemble of these data, it can be seen that it is very difficult to correlate reaction conditions, flow conditions and microreactor characteristics. However, a general observation for all results can be made being that very high yields and conversions can be obtained for the transesterification within just a few minutes in the microreactor, compared with a reaction time of the order of an hour in conventional batch conditions [81]. This can be attributed to the small droplet size and large interfacial area obtained almost immediately in the microreactor.

Although it is not directly clear from the results in Table 4, the characteristic dimension of the microreactor (typically the diameter of the microchannel) has an important effect on the reaction performance as it directly influences drop size and interfacial area. This effect of miniaturisation is clearly illustrated in Table 5. As the microreactor diameter decreases from 2 mm to 250 µm, the interfacial area increases 8-fold. Further, the yield increases whilst the residence time remains fixed, which shows that the increased reaction performance is directly due to the effects of miniaturisation.

Table 5: Influence of the characteristic dimension of the microreactor on interfacial area, a, and on reaction yield. φ: reactor dimension (mm), F: flow-rate (mL/h), a: interfacial area, y: yield (%) [83]

φ (mm)	F (mL/h)	t (min)	a (m³/m²)	y (%)	Reference
2	706.7	8	2000	78.6	[83]
0.53	48.4	8.2	7547	96.7	
0.25	14.7	6	16,000	98.8	

Although reaction performance improves with decreasing drop size, the downside is that phase separation at the outlet of the reactor becomes more difficult if an emulsion is formed. Indeed, smaller drop size typically signifies a more stable dispersion [86]. Nevertheless, it has been shown that the separation of transesterification products is almost instantaneous for slug and parallel flow [83] and [84] in continuous microreactors due to the relatively large size of the slugs/streams compared with droplets in an emulsion formed in batch tanks.

In addition to transesterification reactions, microreactors have also shown to be effective for use in the pretreatment step of WCO via acid-catalyzed esterification for the reduction of FFA. Table 6[87] summarizes the reaction performance of an esterification and a transesterification both carried out with an acid catalyst in a microreactor. The novelty of this work is that the same acid catalyst is employed for both the esterification and the transesterification, which means that the final catalyst neutralization and separation steps are not required. The results show that in the first step of acid-catalyzed esterification, the FFA present in the oil is esterified in 7 min. The second step is the acid-catalyzed transesterification, which has very slow reaction rate in conventional systems (e.g. 90% conversion in 69 h with 1%w sulfuric acid), but demonstrates a dramatic decrease in reaction time in the miniaturized system, with a 99.9% yield obtained in only 5 min [87].

There are no studies on the esterification of fatty acids with glycerol in microreactors reported in the literature. Indeed, operating temperatures above 200 °C are conventionally employed to increase solubility between the reactants. Microreaction technology may therefore be an attractive alternative to current processes since the high interfacial area and the excellent heat transfer characteristics of microreactors could provide mass transfer enhancement with reduced energy consumption. The hydrolysis of triglycerides has however been demonstrated in a microreactor (0.65 mm diameter) using immobilized enzymes at 25 °C. Although the reaction yield is particularly low at this temperature – only 16.8% – it is ten times greater than that obtained in a batch reactor [88]. It can therefore be inferred from these results that the improved reaction performance in the microreactor is due to the miniaturization effects and increased interfacial area between phases.

Table 6: Results obtained for acid-catalyzed esterification in microreactors. y: yield (%), c: conversion (%), R: molar ratio methanol:oil, t: residence time (min), w: weight fraction of catalyst (%), T: temperature (°C), φ : tube diameter (mm), F: flow-rate (mL/h) [87]

Reaction	Oil	Cata-lyst	Performance		Reaction conditions				Type of reactor	Materiau	φ	F	Reference
			y	c	R	t	w	T					
Esterification	Cottonseed (54% FFA)	H_2SO_4	99.1	–	30	7	3	100	Micro-mixer + Tube	Stainless steel	1.2	12	[87]
Transesterification	Cottonseed (0,8% FFA)	H_2SO_4	–	99.9	20	5	3	120	Micro-mixer + Tube	Stain-lesssteel	1.2	12	

Table 7 presents the advantages and drawbacks of microreactors for the different reactions involved in FAME production. In summary, the literature data show that the miniaturization of reactors clearly has a positive effect on mass transfer limited reactions principally due to an increase in the interfacial area between phases. For the best performance of these reactors, it appears important that process parameters be controlled such that an emulsion (i.e. small drop sizes) is formed during the reaction for the enhancement of reaction rates and that slug or parallel flow patterns are formed before the reactor outlet to facilitate the separation of products. The combination of parallel flow and the continuous extraction of one of the products (e.g. glycerol or water, depending on the reaction) also appears as an attractive means to shift reaction equilibrium, as shown by Jachuck et al. [84] for base-catalyzed transesterification. The effects of miniaturization have also shown to provide news ways to perform the reactions, e.g. the use of a common catalyst and reactor for the both the pretreatment of high FFA content feedstock via esterification and FAME production via an acid-catalyzed transesterification, which takes only 5 min compared with tens of hours in conventional equipment. The use of microreactors for mass transfer enhancement in hydrolysis reactions has also been demonstrated and is certainly a means for improving the performance of the esterification of glycerol.

Table 7: Advantages and drawbacks of microreactors for transesterification, esterification with methanol and glycerol, as well as hydrolysis reactions

Microreactors	Advantages	Drawbacks
Transesterification	– short reaction times	– Low flow rates (10–200 mL/)
	– control of the flow pattern to increase mixing during the reaction and to facilitate the separation at the outlet of the reactor	
Esterification (with methanol)	– short reaction times	
	– possibility to keep the same catalyst to transform the remaining glycerides after a pretreatment step	

Esterification (with glycerol)	No studied	
Hydrolysis	– Positive effect	– Yield still low with enzymatic catalysis

The drawback of the use of microchannels and capillary tubes is the low flow rate capacity, which typically ranges from about 10–200 mL/h. However, commercial microreaction technology equipment exist (e.g. reactors by Corning Inc. [89], Chart® [90], Ehrfeld® [91], IMM® [92]) and allow high flow rate capacities up to 10 L/h, which may be more suitable to FAME production at an industrial scale. Indeed, the performance of these reactors for FAME production has rarely been studied and the enhancement of reaction rates and mass transfer due to miniaturization, as well as the ease of product separation are still yet to be demonstrated.

Cavitational Reactors

Cavitation is the generation, growth and collapse of gaseous cavities, which causes the release of large levels of energy in very small volumes, thereby resulting in very high energy densities. The phenomena can occur at millions of locations in the reactor simultaneously, thereby generating conditions of very high local temperature and pressures at overall ambient conditions. The generation of cavities is caused by pressure variations and occurs when the local pressure is less than the saturation pressure. Four techniques exist for the generation of cavitation: acoustic, hydrodynamic, optical and particle. Acoustic and hydrodynamic cavitation is the most commonly employed techniques since the intensity of optic and particle cavitation is insufficient for FAME production. In acoustic cavitation, ultrasounds produce pressure variations in the liquid. Hydrodynamic cavitation, on the other hand, is generated by creating a sudden variation in velocity due to a change in the geometry of the system (e.g. an orifice or venturi). For detailed general information on cavitation and existing technologies, the reader is referred to reviews by Gogate et al. [93], [94] and [95]. When using cavitation to activate liquid phase reactions, two mechanistic steps can be identified. Firstly, the cavity, which contains vapor from the liquid phase or dissolved volatile gases, collapses. The collapsing of the cavity induces extreme temperatures and pressures causing molecules to fragment and generate highly reactive radical species. These species react either within the collapsing bubble or in the bulk liquid. Secondly, the sudden collapse results in an inrush of the liquid, which fills the void and produces shear forces in the surrounding bulk liquid that are capable of breaking the chemical bounds of any molecules. For liquid–liquid reactions, such as transesterification, the cavitational collapse near the liquid–liquid interface causes the rupture of the interface and enhances mixing. This result in very fine emulsions that are typically more stable than those obtained in conventional reactors [93]. Indeed, the creation of such fine emulsions, and therefore high interfacial area between reacting phases, enables mass-transfer limited reactions such as transesterification, esterification or hydrolysis to be greatly enhanced.

Table 8, Table 9 and Table 10 present the performance of esterification, transesterification and hydrolysis reactions activated via acoustic and hydrodynamic cavitation, respectively. Indeed, acoustic cavitation has been more widely studied than hydrodynamic cavitation and detailed reviews of acoustic cavitation as a means for intensifying FAME production have been given in Gole and Gogate [94], Veljkovic et al. [96] and Badday et al. [97]. Ultrasonic cavitation using heterogeneous catalysts have also been reviewed by Ramachandran et al. [98] (see Table 10 and Table 13).

Table 8: Results obtained for transesterification, esterification (with methanol and glycerol) and hydrolysis in acoustic cavitational reactors.y: yield (%), c: conversion (%), R: molar ratio methanol:(acid or oil), t: residence time (min), w: weight fraction of catalyst (%), T: temperature (°C), P: power (W), f: frequency (kHz), V: volume (L), F: flow rate (L/h) [99], [100], [101], [102], [103], [104], [105],[106], [107], [108], [109], [110], [111] and [112]

Oil/acid	Catalyst	Performance		Reaction conditions								Reference
		y	c	R	t	w	T	P	f	V	F	
Acoustic cavitation												
Transesterification (batch)												
Neat vegetable oil	NaOH	98	–	6	20	0.5	25	400	40	–	–	[99]
Vegetable oil	NaOH	98	–	6	20	0.5	36	720 (60%)	40	0.1	–	[100]
Crude cottonseed oil	NaOH	98	–	6.2	8	1	25	–	40	–	–	[101]
Soybean	KOH	99.4	–	6	15	1	40	14.5	20	0.25	–	[102]
Soybean	KOH	99	–	6	5	1	89	400	24	0.125	–	[103]
Soybean	NaOH	95	–	5	1.5	1	40	2200	20	0.75	–	[104]
Soybean	NaOH	98	–	23.9	15	0.5	–	–	–	8	–	[95]
Triolein	KOH	99		6	30	1	ambient	1200	40	–	–	[105]
Beef tallow	KOH	–	92	6	1.17	0.5	60	400	24	2	–	[106]

Continuous transesterification

Canola	KOH	>99	–	5	50	0.7	25		20	0.8	480	[107]
WCO	KOH	1st Reactor: 81	–	2.5	0.53	0.7	ambient	1000	20	0.8	90	[108]
		2nd Reactor: 97.5	–	1.5	0.4	0.3	ambient	–	–	–	120	
		Global: 93.8	–	–	0.93	–	–	–	–	–	–	
Commercial oil	KOH	–	95	6	20	–	40	600	45	2.62	7.8	[109]
		–	~85	6	20	–	40			6.35	19	

Esterification (batch)

Caprylic (C8)	H_2SO_4	–	99	10	75	2	40	120	20	3.5	–	[110]
Capric (C10)	H_2SO_4	–	98	10	75	2	40				–	[110]
Fatty acid	Superacid clay	–	97	10	420	2	40				–	[110]

Esterification with glycerol (batch)

FFA (C8—C10)	H_2SO_4		98.5	3	360	5	90				–	[111]

Hydrolysis

Kerdi oil	None	80		10	600	0	40				–	[112]

Table 9: Results obtained for transesterification in continuous hydrodynamic cavitational reactors. c: conversion (%), R: molar ratio methanol:oil, t: residence time (min), m: molar fraction of catalyst (%), T: temperature (°C), P: pressure drop (bar), (1): first reactor, (2): second reactor [113] and [114]

Oil	Catalyst	Performance	Reaction conditions				Configuration	ΔP		Reference
		c	R	t	m	T		(1)	(2)	
Hydrodynamic cavitation										
Transesterification										
Soybean	NaOH	98.7	6	0.12	3	100	1 Cavitational reactor	17.2	–	[113]
							2 Cavitational reactors in series	17.2	8.3	
Soybean	NaOH	99	6	0.71	3	60	4 Cavitational reactors in series	37.9 (total)		
		99.9	6	0.71	3	100	4 Cavitational reactors in series	37.9 (total)		
Used Frying Oil	KOH	95	–	10	–	60	A plate with 1, 25, 16 or 20 holes	From 1 to 3		[114]

Table 10: Results obtained for esterification with methanol in hydrodynamic cavitational reactors. c: conversion (%), R: molar ratio methanol:oil, glycerol: fatty acid and water: oil, t: residence time (min), w: weight fraction of catalyst (%), T: temperature (°C) [110]

Reactant	Catalyst	Performance	Reaction conditions				Reference
		c	R	t	w	T	
Hydrodynamic cavitation							
Esterification with methanol							
FFA	H_2SO_4	92	10	90	1	30	[110]

In Table 8 it can be seen that the residence time of base-catalyzed transesterification reactions in the presence of ultrasounds is around 15–20 min for most of the cases and only a few minutes or seconds for others. This is significantly shorter than reactions times achieved in conventional reactors that are on the order of one hour. Another interesting point is that short reaction times and high yields can be achieved even at ambient temperature. On the other hand, the reaction time of acid-catalyzed esterification at overall ambient temperatures and pressures is not improved with acoustic cavitation. The heterogeneous superacid clay catalyst used gives long reaction times of about 7 h for a conversion of 97%. Furthermore, the conversion was found to be only 41% when the catalyst is regenerated and reused a second time for a new reaction under the same conditions [110]. The feasibility of glycerol esterification of fatty acids with a homogenous acid catalyst in an ultrasonic reactor has also been demonstrated [111]. Although 98.5% conversion is obtained after a reaction time of 6 h and a temperature of 90 °C, these conditions are better than those required in conventional batch reactors (Gomes and Vergueiro [50] obtain a 90% conversion in 3 h with a molar ratio of fatty acid to glycerol of 3 at 220 °C, and Robles Medina et al. [55] obtain a 96.5% yield in 72 h with an enzymatic catalyst). Indeed, glycerol esterification typically requires high temperature conditions and good mass transfer, and cavitation (via the generation of localized hot spots) appears to be a promising way for enhancing the reaction. The generation of localized hot spots at overall ambient operating temperatures is also the principal interest in performing hydrolysis reactions in cavitation reactors [112]. This reaction type typically requires high temperatures around 200 °C and the limit is set at 250 °C [115] in conventional processes. Cavitational reactors typically allow milder and safer operating conditions, do not require catalysts and enable the reaction time to be reduced (e.g. 10 h to obtain 80% of FFA compared with 12 h to obtain the same quantity of FFA with a solid catalyst in batch conditions [58]).

The most important issue concerning acoustic cavitation is related to reactor scale-up so that large fluid volumes can be processed. The ultrasound probe generates the cavitation phenomena in the vicinity of the probe tip and the major difficulty is then obtaining a homogeneous ultrasonic field throughout the entire reactor volume. Therefore, as the reactor volume increases, an increased amount of ultrasound power must be dissipated to the reaction mixture. A complex design with several powerful probes is required in order to obtain a homogenous acoustic field in larger reactors [96]. Continuous reactors may also be preferred to batch reactors as they enable processing with smaller volumes for the same production capacity. The results obtained by Thanh et al. [108] demonstrate the interest in employing acoustic cavitation in a continuous reactor. Their results show that high yield can be attained in less than a minute at room temperature conditions and a throughput capacity that is potentially of interest for industrial applications. Moreover, Thanh et al. [107] noticed that product separation is facilitated with the decrease of operating temperature and molar ratio of methanol to oil.

Hydrodynamic cavitation is more energy efficient than acoustic cavitation (1×10^{-4}–1×10^{-3} g/J for hydrodynamic cavitation, 5×10^{-6}–2×10^{-5} g/J for acoustic cavitation [110] and [114]) although academic examples of its use for FAME production are relatively rare. Industrial examples of hydrodynamic cavitational reactors are given by Arisdyne Syst. Inc. [116] and Hydro Dynamics Inc. [117], who commercializes the Controlled Flow Cavitation (CFC) and the Shockwave Power Reactor (SPR), respectively. These reactors allow flexible industrial scale flow conditions with capacities ranging from hundreds to tens of hundreds of liters per hour. An example of results obtained for the transesterification of soybean oil using one cavitational reactor or several in series are given in Table 9. Clearly, the main advantage of these reactors is the possibility to obtain almost 100% conversions in only few microseconds. Moreover, an important consequence of such short reaction times is also the reduction of soap formation and emulsification [118], which in turn facilitates the decantation and separation of the products.

The advantages and drawbacks of acoustic and hydrodynamic cavitation for FAME production are summarized in Table 11. From the literature data it can be concluded that transesterification is clearly intensified by cavitation phenomena, which provides short reaction times especially in industrial hydrodynamic cavitational reactors. For the esterification of methanol, however, cavitation does not allow a reduction in reaction times, although it does enable the reaction to be carried out at ambient conditions. Esterification of glycerol is conventionally performed at high temperatures and the use of cavitation allows the operating temperature to be decreased and, at the same time, provides shorter reaction times. Finally, the use of cavitation in hydrolysis of oils has proved to enable milder operating conditions than the usual high temperature and pressure requirements, however, the reaction times still remain long. In terms of flow rate or processing volumes, hydrodynamic cavitation appears to be more adapted to high capacity demands compared with acoustic cavitation devices since obtaining a uniform acoustic field in large volumes is difficult. Indeed, hydrodynamic cavitation has proved to be industrially interesting for the intensification of transesterification reactions; however too few data for the other types of reactions in such reactors are available making it difficult to definitely conclude on the potential of this technique for the intensification of FAME production. Esterification with methanol has been performed in an hydrodynamic cavitational reactor, see Table 10. The performance is similar to esterification in acoustic cavitational reactor.

Table 11: Advantages and drawbacks of acoustic and hydrodynamic cavitation for transesterification, esterification with methanol and glycerol, and hydrolysis reactions

Cavitation	Advantages		Drawbacks	
	Acoustic	Hydrodynamic	Acoustic	Hydrodynamic

Transesterification	– Shorter reaction times than conventional reactors	– Reaction times of only a few microseconds	– Processing of high volumes is difficult due to non-uniform acoustic fields	– A minimum flow rate is required in order to generate the cavitation phenomenon
	– Ambient temperatures and pressures conditions	– Ambient temperature and pressure conditions		
		– Less saponification and emulsion		
		– Less energy consumption than acoustic cavitation		
Esterification (with methanol)	– Ambient temperature conditions		– No particular improvement on reaction rates	
Esterification (with glycerol)	– Milder temperatures than in conventional processes	No data available	– Reaction times remain long	No data available
	– Shorter reaction times than conventional reactors			
Hydrolysis	– Milder operating conditions than in conventional processes	– Milder operating conditions than in conventional processes	– Long reaction times	

Microwave Reactor

The development of microwave technology for the process industries is rather recent and relatively slow because of the lack of control and reproducibility of results, the poor understanding of the dielectric phenomenon occurring, several safety issues and the difficulty to scale-up microwave processes for industrial production. The two major mechanisms involved in microwave technology are dipolar polarization and ionic conduction. Dipolar polarization occurs when dipoles are forced to align with the direction imposed by the electric field, which is caused by the microwave irradiation. The electric field, however, rapidly oscillates and the dipole therefore tries to realign itself with this electric field as fast as possible by rotation. The frequency of microwaves is sufficiently high to cause a phase difference between the field and the dipole orientation and the resulting frictional and collision forces between the molecules thus generate heat. Ionic conduction occurs as the charged dissolved particles oscillate under the influence of the microwave field. When the direction of the electric field is changing, the ions slow down and change direction thereby dissipating kinetic energy as heat. This dissipation is caused by friction [119] and [120]. A more detailed description of microwaves applied in chemistry can be found in [121] and [122]. In addition, two recent reviews focus on biodiesel production assisted by microwaves [123] and [124].

Microwaves are a technology of interest for transesterification, esterification or hydrolysis reactions since they allow increased heating of the reaction medium, which leads to an increase in reaction rate. In early studies on the effects of microwaves in organic synthesis, Lidström et al. [120] observed that for a reaction rate K $= A \cdot \exp(- G/RT)$, the constant A, which describes the molecular mobility, is increased under microwave irradiation due to the increased vibration frequency of the molecules. Terigar et al. [125] also observed that transesterification reaction rates are significantly increased under microwave irradiation and that transesterification with methanol is more sensitive to microwaves than that with ethanol due to the lower gyration radius and molecular inertia of methanol. The efficiency of the transesterification reactions is explained by the dielectric properties of the ionic mixtures and the polar compounds present in the vegetal oil, alcohol and catalyst. Asamuka et al. [126] further attributed the positive effect of microwaves on transesterification to factors other than heating efficiency. Firstly, the conformational isomer of the triglyceride has a lower dipolar moment under microwave irradiation and consequently a lower activation energy. Secondly, the vibration around the $C{=}O$ bond is stronger under microwave irradiation, thereby facilitating the reaction. Finally, the conformational isomer has a planar structure, which is more easily accessible for the nucleophile attack. A comparison of the literature data for homogeneous and heterogeneous base-catalyzed transesterification and for heterogeneous acid-catalyzed esterification is given in Table 12 and Table 14, respectively. From this information, it can be clearly seen that the effect of microwaves on both the transesterification and esterification reactions is a drastic reduction in reaction time (down to several minutes, and even less than a minute in some cases) when compared with conventional processes and without excessive operating temperatures. Indeed, in conventional processes, heterogeneous catalysts for transesterification and esterification are usually associated with low reaction rates and high working temperatures, even though heterogeneous catalysts can facilitate product separation downstream since the solid catalyst is more easily recovered without a neutralization step thereby allowing high purity glycerol as a side product [59]. The literature data also show that under microwave irradiation the performance of heterogeneous catalyzed transesterification reactions – in terms of yield and conversion – are comparable to or better than that when a homogeneous catalyst is used. Indeed, strontium oxide as a solid catalyst gives excellent performance – high conversions for very short residence times – compared with the conventional homogenous potassium hydroxide catalyst [138], see Table 13. In the case of heterogeneous catalyzed esterification, however, the conversions are typically lower than those obtained with a homogeneous catalyst. Indeed, it does appear that the catalyst type has a strong effect since a 100% yield was obtained with scandium triflate, which continues to provide high conversion even after 5 cycles [136], even though process volumes are very low. Further work in the area of catalyst choice is therefore necessary before solid conclusions on the performance of esterification under microwaves can be made. It may also be interesting to test the scandium triflate catalyst in a larger reactor and then develop a continuous process similar to the experimental setup presented by Barnard et al. [135] in order to investigate the industrial potential of this catalyst.

Table 12: Results obtained for homogeneous and heterogenous base-catalyzed transesterification in microwave reactors. y: yield(%), c: conversion (%), c/n: conversion after n cycles, R: molar ratio methanol:oil, t: residence time (min), w: weight fraction of catalyst (%), T: temperature (°C), P: Power (W), V: Volume (mL), F: flow rate (L/h) [127], [128], [129], [130], [131], [132], [133], [134], [135], [136],[137], [138] and [139]

Oil	Catalyst	Performance			Reaction conditions							Reference
		y	c	c/n	R	t	w	T	P	V	F	
Homogenous base-catalyzed transesterification												
Vegetal	NaOH	>97	–	–	–	<2			750	250	–	[127]
Triolein	KOH or NaOH	–	98	–	6	1	5	50	25		–	[128]
WCO	KOH	100	–	–	6	2	1	65	500	500	–	[129]
Rapeseed	KOH	–	93.7	–	6	1	1	40	1200 (67%)		–	[130]
Yellow horn	KOH	>96	–	–	6	6	1	60	500	50	–	[131]
Soybean	NaOH	–	98.64	–	5	20	0.15–0.18	80	1600	270	–	[132]
Rice	NaOH	–	98.82	–	5	20	0.15–0.18	80	1600	270	–	
Safflower	NaOH	98.4	–	–	10	6	1	60	300	500	–	[133]

Feedstock	Catalyst											Reference
Frying palm oil (waste)	NaOH	-	>97	-	12	0.5	3	-	800	16	4.5	[134]
Vegetal	KOH	-	98.9	-	6	0.56	1	50	1600	4000	432	[135]
Heterogeneous base-catalyzed transesterification												
Palmitate	Sc(Otf)3	99	-	97/5	48	2.5	10	20		0.2–2	-	[136]
Yellow horn	$Cs_{2.5}H_{0.5}PW_{12}O_{40}$	96.2	-	96/9	12	0.17	1	60	1000	50	-	[137]
Cooking oil (without FFA)	SrO	-	99	-	6	0.66	1.84	60	900	50	-	[138]
Soybean	SrO	-	-	96/4	6	2	1.84				-	[138]
Cooking oil (without FFA)	SrO	-	99.8	-	-	0.17	1.84		1100		-	[138]
Palm	Eggshell waste (CaO 99.2 wt%)	96.7	-	-	18	4	15	122	900	43	-	[139]

Table 13: Results obtained for transesterification in microwave reactors. R: molar ratio methanol:oil, t: residence time (s), P: Power (W), V: Volume (mL) [138]

Oil	Conversion (%)		Reaction conditions				Reference
	With SrO	With KOH	R	t (s)	P (W)	V (mL)	
Cooking oil (without FFA)	99	83	6	40	900 (70%)	50	[138]
Soybean	96	87	6	120			

Table 14: Results obtained for heterogeneous acid-catalyzed esterification in microwave reactors. y: yield(%), c: conversion (%), c/n: conversion after n cycles, R: molar ratio methanol:oil, t: residence time (min), w: weight fraction of catalyst (%), T: temperature (°C), P: power (W), V: volume (mL) [136], [140], [141], [142], [143] and [144]

Acid	Catalyst	Performance			Reaction conditions						Reference
		y	c	c/n	R	t	w	T	P	V	
Heterogeneous acid-catalyzed esterification with methanol											
Oleic	Sc(Otf)$_3$	100	-	97/5	48	1	1	150	-	0.2–2	[136]
Oleic	NbO$_2$ and ZrO$_2$	-	68	68/3 and 58/3	5	20	10.5	200	1400	80	[140]
Oleic	None	-	60	-	5	60	0	200	1400	80	[141]

Oleic	Dry Amberlyst15	–	39.9	–	20	15	10	60	1600 (100%)	1000	[142]
		–	66.1	–	20	15	10	60	1600 (pulsed 10%)	1000	[142]
Oleic	S_ZrO$_2$	–	>90	–	20	20	5	60	1600	2000	[143]
Heterogeneous acid-catalyzed esterification with glycerol											
Acetic	Starbon-400-SO$_3$H	–	>99	–	1	30	1.8	130	300	<100	[144]

Although glycerol esterification and hydrolysis reactions under microwaves have been rarely studied, the available data suggests that reaction times are typically reduced and less harsh operating conditions are required under microwave irradiation. For example, Luque et al. [144] obtained 99% conversion in 30 min at 130 °C for glycerol esterification of acetic acid using a heterogeneous catalyst (Starbon®-400-SO$_3$H) under microwave irradiation, but only 85% conversion with a sulfuric acid catalyst under the same conditions. Marcel et al. [145] showed that it takes only 5 min to obtain a 100% yield of free fatty acid via the hydrolysis of castor oil triglycerides (in the presence of KOH and ethanol) using microwave irradiation. Another example is given by Saxena et al. [146]: they demonstrated that the complete hydrolysis of triolein with an enzymatic catalyst can be performed in just 75 s at a high power level (800 W, 90 °C), whereas the reaction time needed without at 37 °C and atmospheric conditions is 24 h.

In summary, as presented in Table 15, the current literature shows that microwave irradiation enables a noteworthy reduction of reaction times for transesterification, esterification and hydrolysis and it offers the possibility to operate under milder conditions compared with conventional processes. In addition, the use of microwaves provides a means to obtain good reaction performance with heterogeneous catalysts, which facilitate product separation and higher purity products. Moreover, microwave irradiation is known to break emulsions between a polar and an oil phase [147] thereby facilitating separation further. However, the use of microwave batch reactors for industrial production (i.e. high volumes) is limited for a number of reasons. Firstly, specific security measures have to be taken if microwaves are used at high intensities (power up to 5000 W) that typically require sophisticated cooling systems, thereby increasing the complexity, cost and size of the reactor. Secondly, the penetration depth of microwaves is only few centimeters, which means that a homogeneous field of microwave intensity is extremely difficult to achieve in large volume reactors. For these reasons, most batch experiments in the literature are limited to low volume processing. Indeed, if industrial scale FAME production processes are to be intensified using microwaves then continuous processing will be required and the feasibility of such processes (432 L/h and 98.9% conversion) has already been demonstrated [135]. Furthermore, the combination of microwave-assisted heterogeneous catalysis and continuous processing may have strong industrial potential for FAME production.

Table 15: Advantages and drawbacks of microwave reactors for transesterification, esterification with methanol and glycerol and hydrolysis reactions

Microwaves	Advantages	Drawbacks
Transesterification	– Reaction times reduced	– Scale-up difficult in batch conditions
	– High flow rates for continuous processes	
	– Excellent results with heterogeneous catalysis	
	– Separation of emulsions	
Esterification (with methanol)	– Reaction times reduced	– Studies done with low volumes
	– Good results with heterogeneous catalysis (scandium triflate)	
Esterification (with glycerol)	– Milder temperatures	– Study done with low volumes
	– Good results with heterogeneous catalysis (Starbon-400-SO_3H)	
Hydrolysis	– Short reaction times	– Literature data found with enzymatic catalysis only
	– Effective enzymatic catalysis	

Oscillatory Baffled Reactors

An oscillatory baffled reactor (OBR) is composed of a tube containing equally spaced orifice plate baffles, as shown in Fig. 7a. It operates with an oscillatory or pulsed flow rate, which creates recirculation flow patterns in the vicinity of the baffles as illustrated in Fig. 7b, thereby enhancing mixing and mass and heat transfer. Due to this recirculating flow, OBRs can thereby provide flexible and long residence times, which are comparable to those achieved in batch reactors [148], without having a high length-to-diameter ratio tube. Smaller reactors called mesoscale oscillatory baffled reactor (MOBR) designs with sharp periodic baffles and sharp-edged helical baffles also exist, as presented in Fig. 8.

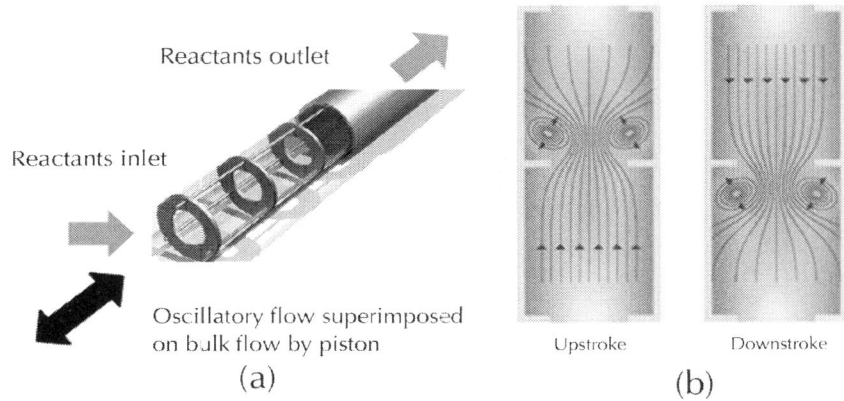

Reactants outlet

Reactants inlet

Oscillatory flow superimposed
on bulk flow by piston

(a)

Upstroke   Downstroke

(b)

Figure 7: The oscillatory baffled reactor (a) and the flow-pattern corresponding to a upstroke or downstroke current (b) [150].

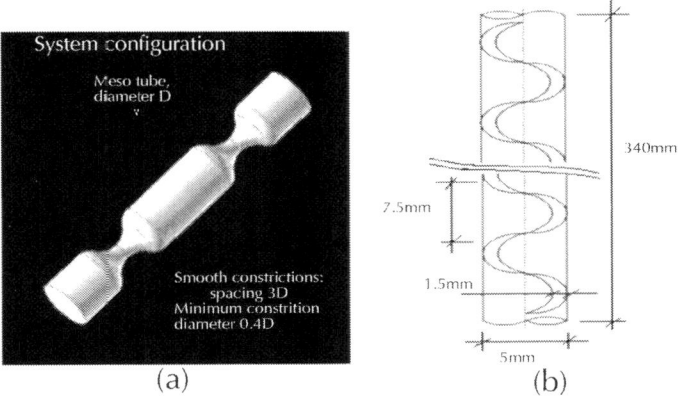

System configuration

Meso tube,
diameter D

Smooth constrictions:
spacing 3D
Minimum constrition
diameter 0.4D

(a)

340mm

7.5mm

1.5mm

5mm

(b)

Figure 8: Mesoscale oscillatory baffled reactors with (a) sharp periodic baffles (SPB) [150] or (b) sharp-edged helical baffles (SEHB) [149].

The OBR technology is particularly adapted to liquid–liquid reactions, such as transesterification, because it allows good inter-phase contacting, enhanced mixing and sufficiently long residence times for reaction. Table 16 compares FAME production data obtained in OBRs; note that the literature studies have been limited to transesterification only. The results show that in an OBR, 99% conversion can be reached in only 10 min at 60 °C, which is half the time required to carry out the reaction in conventional reactors [148]. Furthermore, the same study shows that the methanol stream can be recycled in the continuous OBR technology, thereby allowing a very low methanol to oil ratio to be used. Mesoscale OBRs, which are particularly adapted to feasibility studies and screening tests due to their lower capacity, have also been shown to provide high conversions and yields in short times compared with what can be achieved in conventional batch tanks [149] and [150].

Table 16: Results obtained for transesterification in oscillatory baffled reactors. y: yield(%), c: conversion (%), R: molar ratio methanol:oil, t: residence time (min), T: temperature (°C), V: volume (mL), F: flow rate (L/h) [148], [149] and [150]

Oil	Catalyst	Perfor-mance		Reaction conditions					Type of reactor	Refer-ence
		y	c	R	t	T	V	F		
Rape-seed	NaOH	–	99	1.5	30	50	1.56	3.12	OBR	[148]
Refined vegeta-ble oil	NaOMe	–	99	6	40	60	0.005	0.126	SPB-MOBR	[150]
Rape-seed	KOH	90	–	9	20–25	50	0.04	0.12	SEHB-MOBR	[149]

In addition to the advantages of OBRs for reaction performance, which are summarized in Table 17, this type of equipment is particularly suited to industrial scale production where a certain flow capacity may be required. Indeed, commercial equipment exists – such as that developed by NiTech Solutions®, which has demonstrated successful biodiesel production [151]. OBRs have also shown to provide good solids handling, whether it be solids suspension or crystallization applications [148] and [151], which is interesting if heterogeneous catalysis is to be used. Although only transesterification reactions in OBRs have been demonstrated in the literature, the results of these studies and the characteristics of this continuous process equipment suggest that they could also provide significant advantages for esterification and hydrolysis reactions.

Table 17: Advantages and drawbacks of OBRs for transesterification reaction

OBR	Advantages	Drawbacks
Transesterification	– Reaction times significantly reduced compared with batch processing	– No integrated separation unit for end products
	– Compatible with heterogeneous catalysis	
	– Molar ratio methanol to oil reduced	
	– Large flexibility in residence times	

Static Mixers and Other Motionless Inline Device

Static mixers are motionless elements that are inserted in a tube or pipe and enable fluid mixing by creating transverse flows. They are typically used in continuous processes but can also be employed in a closed loop system or for premixing before feeding to a batch tank. Static mixers are well adapted to variety of applications, including simple blending and multiphase mixing problems, in both the laminar and turbulent flow regimes [152]. The advantages of static mixers over conventional batch stirred tanks are their smaller size, lower energy consumption, a very good control of the residence time with a plug flow reactor behavior, and finally very good mixing with low shear rates [153]. A large number of more or less intricate static mixer designs exist – about 2000 US patents have been granted and more than 30 models are commercially available, e.g. Kenics, HEV, KMS and KMX mixers by Chemineer Inc., Dayton, OH (see Fig. 9) and the SMX and SMX plus mixers by Sulzer Chemtech, Switzerland. However, other inline devices, such as randomly packed beds of spherical particles and metallic foams, can also be used to ensure mixing in tubes and pipes with relatively low pressure drop [154] and [155].

HEV Mixer   KM Mixer   KMX Mixer

Figure 9: Geometries of commercially available mixers [156].

Amongst the different applications, static mixers are well adapted to liquid–liquid dispersion processes, including extraction, reaction and emulsification, providing a means to disperse immiscible phases with typically less energy requirements than other technologies. This has been demonstrated by Frascari et al.[157] who have shown that in the case of a transesterification reaction, the energy consumption for a static mixer is less than that required by mechanical stirring. In particular, they show that if the reactants are premixed using a static mixer and fed to a batch reactor with two 4-bladed disk turbines rotating at 100 rpm, the conversion is the same as that achieved in the batch reactor alone with an impeller speed of 700 rpm. Further, they show that the droplet size created by the static mixer is comparable to that generated with mechanical agitation at 700 rpm, which explains the similar reaction performance.

The studies reported in the literature concerning the transesterification of vegetable oil mostly employ static or inline mixing devices in a continuous mode. The transesterification reaction performance obtained in different static mixer and inline mixing configurations are summarized in Table 18. Thompson and He [158] developed a static mixer closed-loop system, which allows the residence time to be easily varied. With this system, the production of total free glycerol (ASTM D6584 specification, 0.24%w max.) was achieved in a residence time of 15 min. Boucher et al. [159] designed a system with a static mixer and an integrated decanter as shown in Fig. 10. The immiscible reactants are contacted in the static mixer and the products are released into a glass column. Where the glycerol settles to the bottom and is continuously removed, whilst the esters are continuously removed from the top of the column. The same group then improved the transesterification process performance by inclining the reactor as presented in Fig. 11[160]. By inclining the reactor, larger glycerol droplets are formed, thereby facilitating the separation of the esters and glycerol, although the conversion is slightly affected with a decrease from 99% to 95% probably due to the increased drop size. In both reactor designs, the phase contacting via the static mixer and the continual removal of glycerol enhance the reaction and high conversion is obtained in less than 30 min, which is significantly faster than in conventional equipment. Metallic foams and other inserts, such as steel spheres and various types of fibers, have also shown to enhance transesterification reactions [155], [161], [162] and [163]. These devices create tortuous interstices with micron-sized dimensions that promote fluid contacting and mixing, and consequently the chemical reaction. The data show that these mixing devices allow high conversion rates in short residence times, which are of the order of several minutes. Furthermore, some authors also observed good product separation at the reactor outlet, despite the formation of liquid–liquid emulsions [163].

Table 18: Results obtained for transesterification in static mixer reactors. y: yield(%), c: conversion (%), g: glycerol treatment, R: molar ratio methanol:oil, t: residence time (min), w: weight fraction of catalyst (%), T: temperature (°C), F: flow rate (L/h) [158], [159], [160],[161], [162] and [163]

Oil	Catalyst	Performance			Reaction conditions					Type of reactor	Reference
		y	c	g	R	t	w	T	F		
Canola	NaOH		–	–	6	30	1.5	60	0.104	Closed-loop system	[158]
Waste canola (pretreated)	KOH		99	70–99% removal	6	19	1.3	40–50	72	Reactor-separator	[159]
Waste oil	KOH		96	36–95% separation	6	17.5	1–1.3	40–50	72	Reactor-separator	[160]
Soybean	NaOH	95.2	–	–	10	2.16	1	55	0.918	Metal foam	[155]
Soybean	KOH	98.2	–	–	6	0.99	2	60	0.366	Tube filled with stainless steel spheres (2.5 and 1.0 mm)	[161]
	KOH	97.05	–	–	6	3	1	60	0.279	Packed bed reactor with 2.5 mm spheres	[162]
Soybean	NaOMe	–	99	–	–	5.8	2	60	1.44	216,000 Fibers 8 μm	[163]

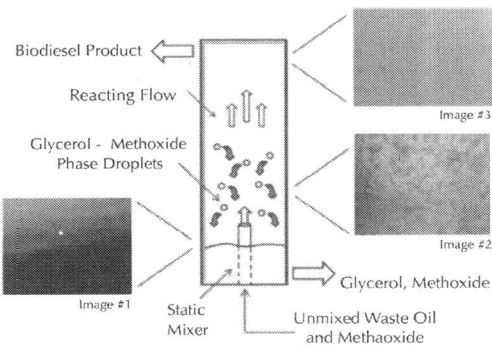

Figure 10: Schematic diagram of the static mixer reactor/separator and corresponding images of the flow at different positions within the reactor [159].

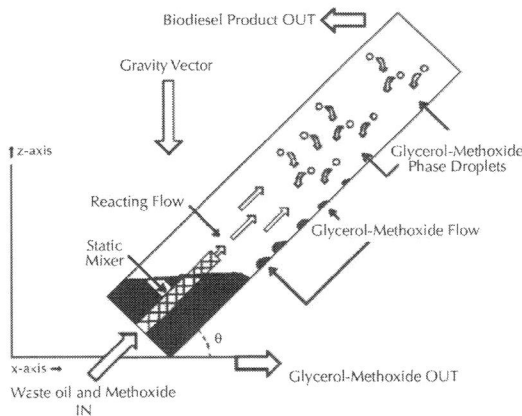

Figure 11: Schematic diagram of the inclined static mixer reactor/separator and operating concept [160].

A summary of the advantages and drawbacks of static mixers and inline devices is given in Table 19. In general, this type of equipment allows high reaction conversions to be reached in relatively short times and requires less energy input compared with conventional batch processing. A novel attribute of these devices is the possibility to couple the mixing/reaction and separation steps that are required for the transesterification process. Continual separation and removal of products during the reaction shifts the equilibrium and means that the sequential product separation steps that are required in conventional processes can be avoided. Indeed, the other reactions related to FAME production, i.e. esterification with various alcohols and hydrolysis could potentially benefit from the integration of the reaction and separation steps in one device, although no studies have yet been dedicated to this. Also, heterogeneous catalysis has not yet been tested with static mixing and inline technologies; indeed solids handling in some of these devices may be difficult (e.g. due to clogging), however catalytic coatings on foams or fibers may be a means to resolve such problems.

Table 19: Advantages and drawbacks of static mixers for the reaction of transesterification

Static mixers	Advantages	Drawbacks
Transesterification	– Reaction times reduced (×2–10)	– May not be adapted to heterogeneous catalysis
	– Low energy-consumption	
	– Relatively simple devices	
	– Separation step implemented in the reactive part for the reactor-decanter	

Membrane Reactors

Microporous inorganic membranes enable product separation by a molecular sieving effect. They can be made of ceramic, zeolites, silica, carbon or polymers. Carbon membranes are most commonly used due to their easy production and low cost [164]. Membrane reactors are also employed for pervaporation processes that enable the separation of a liquid retentate from a vapor permeate [165]. A detailed review of membrane technology as an alternative means for biodiesel production has been given by Shuit et al. [166]. In the case of transesterification, the principal objective of the membrane reactor is to retain the triglycerides [167], [168], [169], [170], [171] and [172]; this then facilitates the downstream purification steps [167]. The retention of mono- di-, and triglycerides is also possible [168], thus allowing the separation of FAME and glycerol at room temperature and avoiding onerous downstream processing. Furthermore, the separation of glycerol reduces waste water generated by washing steps [171]. The retention of soaps and glycerol is also possible by adding a little amount of water as depicted in Fig. 12[172]. Membrane separation also allows product specifications to be met (i.e. glycerol < 0.2%w in FAME), as well as the treatment of waste cooking oils with high FFA content (5%) [173]. as it is summarized by Sdrula [170], others advantages of membrane reactors for transesterification are the high purification of glycerol, the absence of chemical additives and the low process cost.

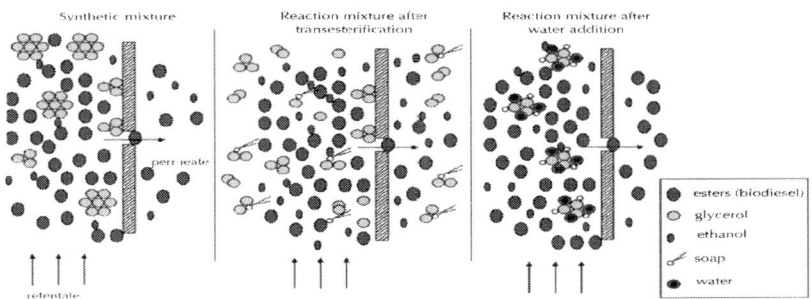

Figure 12: Mechanism of separation of glycerol by microfiltration with a ceramic membrane [172].

Numerous studies have shown that membranes reactors can improve conversion and facilitate the product purification step of transesterification reactions, compared with conventional processes [172]. A summary of the performance of transesterification reactions carried out in membrane reactors is given in Table 20. It can be seen that in most cases good conversion and acceptable flow rates can be achieved, however the time required for the reaction is on the order of 1–2 h, which is similar to conventional batch processing.

Table 20: Results obtained for transesterification in membrane reactors. y: yield(%), c: conversion (%), g: glycerol content in permeate (%), R: molar ratio methanol:oil, t: residence time (min), w: weight fraction of catalyst (%), T: temperature (°C), F: flow rate. φ: pore size (μm), V: volume of the reactor (L) [167], [168], [169], [171], [172] and [173]

Oil	Catalyst	Performance			Reaction conditions					Reactor			Reference
		y	c	g	R	t	w	T	F	Membrane	φ	V	
Canola	H_2SO_4	64	-	-	-	49	6	70	16.6 L/m²/h	Carbon	0.05	0.3	[167]
	NaOH	96	-	-	-	49	1	70	16.6 L/m²/h	Carbon	0.05	0.3	
Canola	NaOH	98.7	-	0	16	210	0.5	-	9 L/m²/h	Carbon	0.2	0.32	[168]
Canola	NaOH	98	-	-	24	5	0.5	65	120 kg/h	Filtanium ceramic	-	6	[169]
Canola	NaOH	-	-	<0.02	6	180	1	25	40–180 L/m²/h	Polyacrylonitrile membrane	-	-	[171]
Soybean (Ethanol)	NaOH	-	98.7	<0.02	9	80	1	30	10.3 kg/m²/h	Ceramic	0.2	0.25	[172]
WCO (FFA = 5%)	Base	>99	-	<0.02	23.9	35–105	0.5–1.4	65	30–40 L/m²/h	Titanium oxide	0.03	0.48–0.63–0.74	[173]

Esterification has been demonstrated in membrane reactors via a pervaporation process where water is removed in a vapor stream allowing a shift in the reaction equilibrium, thereby leading to higher conversion rates. The results obtain by Sarkar et al. [174], which are given in Table 21; show that almost 100% conversion of the fatty acid is achieved in 6 h.

Table 21: Results obtained for esterification in membrane reactors. y: yield(%), c: conversion (%), g: glycerol content in permeate (%), R: molar ratio methanol:oil, t: residence time (min), w: weight fraction of catalyst (%), T: temperature (°C), t: residence time (h) [174]

Acid	Catalyst	Performance		Reaction conditions				Reactor			Reference
		y	c	R	w	T	t	Membrane	(μm)	V (L)	
Oleic	H_2SO_4	–	99.9	27	0.3	65	6	Polyvinylalcohol on polyether sulfone	–	0.073	[174]

The advantages and drawbacks of membrane reactors for FAME production processes are summarizedTable 22. The major advantage of this technology is that separation is integrated in the reaction step. This in turn allows high reaction conversion to be achieved; however there is no recuction in reaction time compared with conventional processing. Although no studies have yet focused on the hydrolysis of triglycerides or esterification using glycerol, membrane reactors could potentially provide benefits by shifting the reaction equilibrium, thereby promoting product formation and facilitating the product separation. Heterogeneous catalysts can be incorporated in and mobilized on the membranes, leading to conversions superior to 90% without washing steps [175], [176] and [177]. The use of membrane reactors with an immobilized enzymatic catalyst for hydrolysis has also been demonstrated [178].

Table 22: Advantages and drawbacks of membrane reactors for the reaction of transesterification

Membrane reactors	Advantages	Drawbacks
Transesterification	– Retains Triglycerides (unreacted glycerides stay in the reactor), economy of downstream process costs	– No increase of reaction rates
	– Retains Di and Monoglycerides, allowing phase separation at room temperature	
	– Better separation of glycerol means less washing and less waste water	
	– Retains Glycerol, separation step during the reaction	
	– Low cost	
Esterification	– Separation of water during the reaction: shift of the equilibrium	– No heterogeneous catalysts used

Reactive Distillation

Reactive distillation combines reaction and distillation in a multifunctional reactor and is suited to heterogeneous, homogeneous and non-catalyzed reactions [179]. The principle is based on the removal of one reaction product in order to shift the reaction equilibrium, thereby leading to high and even total conversions. A more complex configuration is the reactive divided-wall column, which enables three high-purity streams to be obtained at the outlet of a single distillation tower as depicted in Fig. 13[180]. In conventional distillation, two distillation columns are required to separate three products.

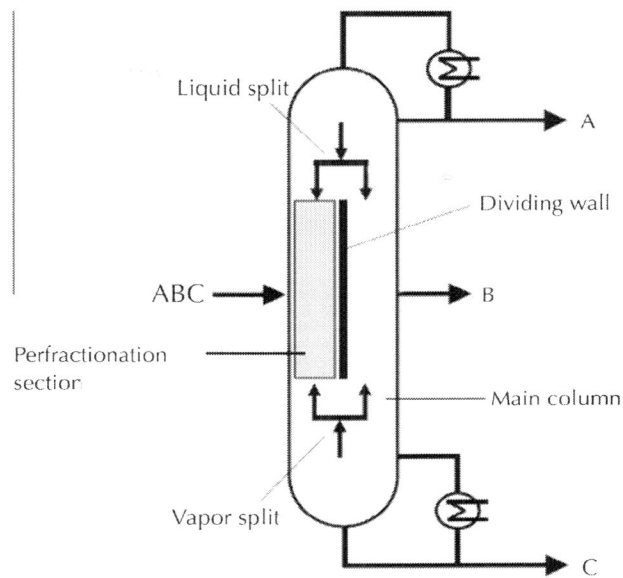

Figure 13: Dividing wall column [180].

Reactive distillation has shown to be particularly adapted to transesterification and esterification reactions since these are equilibrium reactions that benefit from equilibrium shifts. The literature data on the performance of transesterification and esterification reactions are presented in Table 23 and Table 24, respectively.

Table 23: Results obtained for transesterification in a reactive distillation column. y: yield(%), c: conversion (%), R: molar ratio methanol:oil, t: residence time (min), w: weight fraction of catalyst (%), T: temperature (°C), F: flow rate (L/h) [181]

Oil	Catalyst	Reaction conditions						Reference
		y	c	R	t	w	F	
Canola	KOH	94.4	95	4	3	1	4.9 L/h	[181]

Table 24: Results obtained for esterification in a reactive distillation column. y: yield(%), c: conversion (%), R: molar ratio methanol:oil, t: residence time (min), T: temperature (°C), P: pressure (bar), F: flow rate [182] and [183]

Acid	Catalyst	Reaction conditions							Reference
		y	c	R	t	T	P	F (mol/h)	
Decanoic	Amberlyst 15		100	2	–	50	3	34 mol/h	[182]
Dodecanoic			>99.99	1				1250 kg/h	[183]

In the case of reactive distillation for transesterification reactions, the reaction of methanol and triglycerides and the separation of any excess or unreacted methanol are achieved simultaneously [181] and [184]. Unreacted methanol is recovered at the top of the column and is recycled to the feedstock, whilst the glycerol and esters are recovered at the bottom of the column and sent to a decanter. This processing method enables the use of a lower methanol to oil molar ratio and provides short residence times.

Esterification has also been successfully carried out by reactive distillation. It provides the opportunity to use a heterogeneous catalyst, thereby avoiding the neutralization, washing, separation and waste recovery steps [185]. It also enables the removal of water throughout the reaction at the top of the column, which shifts the reaction equilibrium and promotes product formation. Steinigeweg and Gmehling [182] used Amberlyst 15, an ion exchange resin that is fixed on the packing, to esterify decanoic acid and obtained 100% conversion at a pilot scale with a flow rate of 34 mol/h. Kiss et al. [185] investigated the choice of the heterogeneous catalyst. They found that zeolites have too small pores, which limit the diffusion of large molecules like fatty acids or esters. Ion exchange resins, such as Nafion or Amberlyst, have strong acid activity but a low thermal stability. Tungstophosphoric acid is very active but is also soluble in water and therefore cannot be reused. The authors concluded that sulfated zirconia is a good candidate as catalyst for esterification because it is active, stable and selective. Dimian et al. [186] investigated the possibility of using a co-solvent – 2-ethylhexanol – to increase the immiscibility between water and fatty acids so that the water can be more easily separated.

Fig. 14 shows the concept of reactive absorption using a heterogeneous catalyst as proposed by Kiss and Bildea [187]. Fatty acids and methanol are fed at the top and bottom of the column, respectively, and water with some acid exits at the top of the column, whilst the bottom outlet stream contains FAME with some methanol.

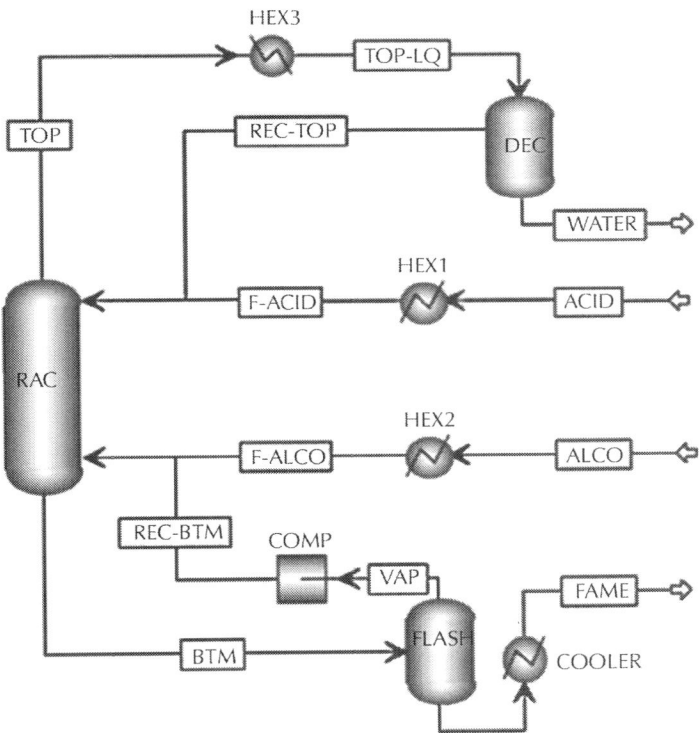

Figure 14: Diagram of the reactive absorption. Water is recovered at the top of the column, FAME is recovered at the bottom of the column [187].

The reactive distillation process can be further improved by using a divided wall column (DWC), which is particularly useful for reactive systems that have more than two products or that operate with an excess of reagent. This technology, which is analogous to the assembly of two distillation columns in one unit, allows the separation of multicomponents [188] and solves the problem encountered in reactive absorption, which is the necessity to use methanol in an exact stoichiometric ratio since it has to be completely converted. Indeed, this ratio is difficult to obtain especially because the amount of fatty acids in the feed is often unknown. Reactive distillation using a DWC allows the use of an excess of methanol, which is then recovered as the top distillate; water is then recovered as a side stream and FAME as the bottom product, as depicted in Fig. 15[183].

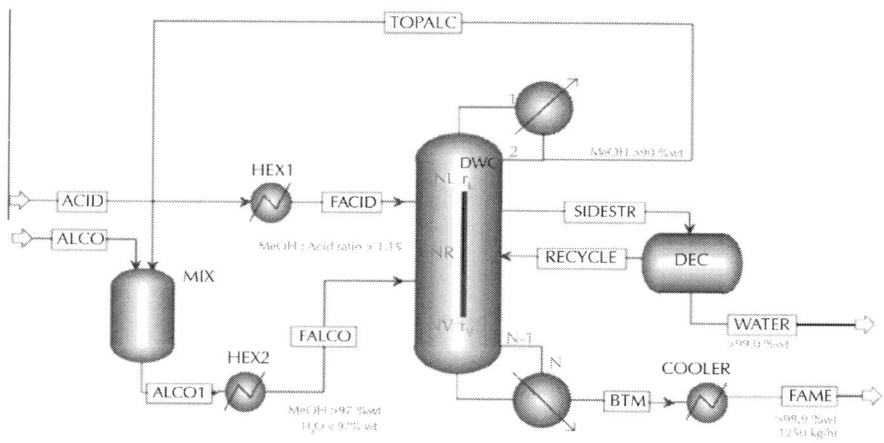

Figure 15: Reactive dividing-wall column configuration [183]. Excess of methanol is recycled at the top, water is removed at the side and FAME is recovered at the bottom.

Although there have been no demonstrative studies on the use of reactive distillation for the esterification with glycerol presented in the literature, this technology may be of interest for the reaction due to the high boiling points of glycerol and fatty acids, thereby allowing the removal of methanol and water. The use of reactive distillation for hydroesterification (i.e. combination of hydrolysis and esterification is also of potential interest. Indeed, hydrolysis can be firstly performed with reactive distillation to concentrate the FFA content of oils that already have a high level of FFA before carrying out the esterification. This has been demonstrated numerically in a two-section reactive distillation process using WCO and Jatropha oil [189].

The advantages and drawbacks of reactive distillation for FAME production are summarized in Table 25. The main advantage of reactive distillation for FAME reactions is the fact that reaction and separation can be combined in the same device, thereby enabling the reaction equilibrium to be shifted and enhancing product formation. The principal disadvantage of this type of process is the high energy requirement, which is also an issue in conventional distillation processes.

Table 25: Advantages and drawbacks of reactive distillation columns for the reaction of transesterification, esterification and hydrolysis

Reactive Distillation	Advantages	Drawbacks
Transesterification	– Reaction and separation of the methanol in one single step	– Energy requirements
Esterification (with methanol)	– Reaction and separation in one single step	– Energy requirements (10^{-3} $g_{FAME}/J_{reboiler}$)
	– Possibility to have 3 streams, methanol (top), water (middle) and FAME (bottom) with R-DWC	(136 kW at the reboiler for a production of 1250 kg/h, data of Table 3 from Ref. [187])
	– Compatible with a hetero-geneous catalyst	
Esterification (with glycerol)	– Remove of water (with methanol)	– No results found
Hydrolysis		– Products with high boiling points

RECOMMENDATIONS FOR THE CHOICE OF PROCESS TECHNOLO- GIES FOR FAME PRODUCTION

Mass-transfer is the main limitation of the WCO transformation reactions, however, two additional parameters are also important when considering the choice of process technology. Firstly, the choice of the catalyst directly impacts the process performance. Secondly, the possibility to combine the reaction and separation steps to shift reaction equilibriums and easily separate products is important for process efficiency. The objective of this section is to give recommendations, following the review of literature data, on the choice of equipment taking into account the catalyst type and the separation aspects for the intensification of the different transesterification, esterification and hydrolysis reactions.

Catalyst Type

Table 26 summarizes the different types of catalysts that have been used for FAME production in various types of innovative process equipment. It can be seen that majority of studies have employed homogenous catalysts for transesterification and esterification reactions because they are inexpensive and easy to use. On the other hand, heterogeneous catalysts have been less widely used due to traditionally higher operating temperatures and long reaction times. However, new catalysts, such as strontium oxide, scandium triflate or sulfated zirconia, have been developed and give similar performance to homogenous catalysts. For sustainable FAME production processes, the ensemble of the literature studies suggest that heterogeneous catalysts may be preferred over the conventional homogeneous catalysts because they can be easily reused and regenerated, and also allow easy product separation and elimination of the neutralization step. Moreover, heterogeneous catalysts coupled with intensified processes, such as microwaves and reactive distillation, have been shown to significantly enhance FAME production[136] and [183]. It is expected that heterogeneous catalysts may also provide improved reaction performance in other types of innovative continuous flow equipment, e.g. microstructured reactors, cavitational reactors, OBRs and static mixers, which have been proven successful for other applications in solids handling (suspensions or catalytic reactions) [110], [190], [191] and [192]. Indeed, further exploratory work on the use of heterogeneous catalysts in innovative process equipment for transesterification or other FAME production reactions is still required.

Table 26: Summary of the type of catalyst used and the ease of separation in different process technologies for the four reactions.+ indicates a positive effect of the technology; – indicates a negative effect; 0 indicates no effect. The data without brackets has been proven in literature studies whereas data between brackets is deduced from the results in the literature

Reactor	Catalysis			Hydrolysis	Separation facilitation			
	Trans-esterification	Esterification			Trans-esterification	Esterification		Hydrolysis
		Methanol	Glycerol			Methanol	Glycerol	
Microreactor	Homogenous base	Homogenous acid	Enzymes	Enzymes	++	(++)	(++)	(++)
	Homogenous acid							
Cavitation	Homogenous base	Homogenous acid	Homogenous acid	None	+	(0)	(0)	(0)
		Heterogeneous acid						
Microwave	Homogenous base	Homogenous acid	Enzymes	Homogenous base	++	++	++	++
	Heterogeneous base	Heterogeneous acid	Heterogeneous acid	Enzymes				

Oscil-latory Baffled	Homog-enous base	Homog-enous acid	/	/	0	(0)	(0)	(0)
Static mixers	Homog-enous base	Homog-enous acid	/	/	0	(0)	(0)	(0)
Mem-brane reactor	Homog-enous base	Homog-enous acid	/	/	+++	+++	(+++)	(+++)
Reactive distilla-tion	Homog-enous base	Homog-enous acid	Homog-enous acid	None	+++	+++	(+++)	(– – –)
		Heteroge-neous acid						

Product Separation

Product separation is an important point to be considered in the design of sustainable processes since separation steps can be highly costly in terms of energy and time [38]. Separation steps can be combined with mixing and reaction operations in innovative equipment thereby shifting reaction equilibrium and enhancing process performance. Product separation and purification can also be facilitated by the use of heterogeneous catalysis and by inducing specific physical or chemical phenomena in the different process equipment.

Table 26 indicates how easy it is to separate products for the four reactions with different types of catalysts in various process equipment. Results from the literature show that microstructured reactors accelerate decantation due to the formation of parallel flow patterns or of bigger droplets with slug-flow. In cavitational reactors, the reaction is very fast, thereby hindering the formation of di and monoglycerides, which have surfactant properties. As a result, the emulsion is unstable and the phases are easy to separate. Microwave irradiation has also shown to accelerate the decantation process by enhancing drop coalescence. Membrane reactors have been shown to retain glycerides and soaps in transesterification, leading to highly pure products. Pervaporation membrane is also used to remove continuously water during esterification. It is expected that this type of technology may also be of interest for the other reactions. Reactive distillation has of course proven to be adapted to product separation for transesterification and esterification. However, it may not be of interest for the hydrolysis of triglycerides because it would remove water, which is in this case a reactant, and would shift equilibrium in the wrong direction. Oscillatory baffled reactors and static mixers do not integrate any specific separation operation. As a result, they most often followed by a decanter for phase separation. However, novel devices such as static mixer with integrated decanter have shown to shift the transesterification reaction equilibrium by withdrawing the glycerol product [159] and [160].

Equipment Choice

Table 27 presents recommended choices of catalyst type and process equipment, for both reaction and separation, for the four different reactions related to FAME production and provides characteristic information on reaction conversion, time and the process energy requirements.

Table 27: Summary of heterogeneous catalyzed reactions with forces of the different technologies. Technologies highlighted in bold are technologies of interest on an energy consumption criterion. c: conversion (%), t: residence time (min)

	Catalysis	Process Equipment					
		Reactor performance		Recommended equipment for mass transfer	Selection criteria	Recommended equipment for product separation	Energy requirements (g/l)
		c(%)	t(min)				
Transesterification	Heterogeneous (SrO/SiO2) FFA and water tolerance increase to 3%	96–99	2–30	Static mixers	Less energy consuming	Decanter: glycerol removal	10^1
		97–100	1–6	Millireactors	Reaction time reduced	Flow-pattern: accelerated decantation	10^{-2}
		95–99	<1 μs	Cavitational reactors		No DG/MG formation	10^{-4}
		96–99.8	1–6	Microwave		Break emulsion between a polar and an oil phase	10^1
		96–99	5–210	Membrane	Integrated separation unit	Membrane: Keep unreacted glycerides (higher purity)	/
Esterification (Methanol)	Heterogeneous (Sulfated zirconia, Scandium trisulfate)	100	/	Reactive distillation	Shifted equilibrium due to product removal	Separation of water, methanol and esters	10^{-6}
		/	/	Static mixers		Decanter (water removal)	10^1
		99.9	/	Membrane		Membrane: Removal of water	/

Esterification (Glycerol)	Heterogeneous (Sulfated zirconia, Scandium trisulfate)	/	/	Cavitational	Mixing intensification (reduce T, P)	Compatibility with an integrated decanter	10^{-4}
		/	/	Millireactors	Reaction time reduced	Recovery of unreacted glycerol	10^{-2}
		/	/	OBR			–
		99	30	Microwave		Break emulsion between a polar and an oil phase	10^{1}
		/	/	Reactive distillation	Shifted equilibrium due to product removal	Removal of water	$>10^{-6}$
Hydrolysis	Heterogeneous (SrO/SiO$_2$) (FFA and water tolerance increase to 3%)	/	/	Cavitational	Mixing intensification (reduce T, P)	Compatibility with an integrated decanter	10^{-4}
		/	/	Millireactors		Recovery of unreacted glycerol	$>10^{-6}$
		100	5	Microwave		Distillation (under vacuum)	10^{-2}
		/	/	Membrane	Integrated separation unit	Membrane:Keep unreacted glycerides (higher purity) retains soaps	/

Transesterification

Transesterification is the most common reaction used for obtaining FAME. In terms of catalysts, strontium oxide with silica is a very good candidate for transesterification reactions because its tolerance to FFA and water is high compared with other catalysts. Indeed, conversion of more than 90% can be achieved with FFA and water contents greater than 3 wt. %. Under these conditions, this means that WCO pretreatment steps may be not always be essential.

In transesterification reactions, the mixing process is important at the beginning of the operation where the objective is to increase the interfacial area between the triglycerides and the alcohol so that the reaction can take place. However, once the reaction starts, the generation of interfacial area by mixing is less difficult due to the formation of di- and monoglycerides, which facilitate liquid–liquid contacting and dispersion, leading to a pseudo-homogenous phase. Therefore, sophisticated mixing technology is not particularly required, which is why a low energy consumption reactor with static mixers may often be preferred. However, if the process objective is to reduce processing times, other intensified reactors such cavitational, microstructured, microwave or oscillatory baffled reactors are recommended. In such equipment, the separation step of the transesterification process is facilitated naturally by inherent characteristics of the process and flow conditions or by combining it with a specific device for separation. For example: micro/microstructured reactors enable the generation of liquid–liquid flow patterns that ease product separation at the outlet; the formation of di- and monoglycerides and consequent pseudo-homogenous phase is avoided in cavitational reactors, thereby leading to a less stable emulsion and facilitating phase separation; microwaves promote the drop coalescence process, thereby speeding up the separation process. These different equipment types can also be used in combination with a decanter for product separation or with a membrane, which retains any unreacted glycerides and allows high purity esters to be obtained at 90% conversion.

Esterification with Methanol

In FAME production processes, esterification with methanol is often performed to reduce FFA levels in feedstock before performing base-catalyzed transesterification. Heterogeneous catalysts are preferred for the same reasons than previously. Good catalyst candidates are sulfated zirconia or scandium trisulfate, which have demonstrated high performance in reactive distillation and microwave reactors, respectively.

In esterification reactions, the mixing process has a more or less important role depending on the miscibility of the reactants, which is related to the length of the carbon chains. For immiscible reactants, mixing must be able to generate sufficient interfacial area between the phases and flow circulation to promote the reaction. In the case of partially miscible reactants, drop break-up is less important but the mixing of reactants is still vital for reaction performance. Amongst the different innovative process equipment, both reactive distillation and static mixer systems with an integrated decanter appear to give best results for esterification. Indeed, in these systems the products are continuously separated from the reactive media, thereby shifting the reaction equilibrium and enhancing product formation. The use of membranes could also be considered in order to retain unreacted glycerides and produce high purity esters.

Esterification with Glycerol

Esterification using glycerol, which is a side-product of transesterification, is an interesting way to decrease the FFA levels in high FFA content WCO. Heterogeneous acid catalysis using sulfated zirconia or scandium triflate appears to be the most appropriate means to catalyse the reaction.

In esterification reactions using glycerol, mixing is an important step of the process due to the very low solubility of reactants (i.e. fatty acids and glycerol). Indeed, the mixing operation determines interfacial area available for mass transfer and the fluid circulation, both of which are important for the progress of the reaction. Conventionally, high temperatures are used to increase reactant solubility and to facilitate fluid contacting with the disadvantage of high operating costs. Cavitational, microwave, oscillatory baffled and microstructured reactors provide a means to intensify mixing and fluid contacting at ambient operating conditions, thereby being more energy efficient. Reactive distillation under vacuum conditions can also be used to shift the reaction equilibrium and separate products, but with the disadvantage of requiring higher energy input. A static mixer reactor-decanter that shifts the reaction equilibrium may also be a recommended choice if reaction yield is the principal process objective.

Hydrolysis

The primary interest of the hydrolysis of triglycerides is to concentrate fatty acids in the feedstock. Strontium oxide is possibly a good candidate for heterogeneous catalysis. Like for the esterification reaction with glycerol, the mixing process is very important due to low solubility of reactants. As a result, conventional processing often requires high temperatures and pressures to ensure sufficient yield and acceptable reaction times. Intensified process technologies such as cavitational, microwave, oscillatory baffled or microstructure reactors can be used to intensify mixing and allow good reaction performance with milder operating conditions. Reactive distillation is not recommended since it would result in the removal of water and would not promote the formation of products. The separation of products – fatty acids and glycerol – can be carried out with a distillation column under vacuum conditions or in microstructured reactors by generating slug-flow or parallel flow, which accelerates the decanting process. Since glycerol and fatty acids are miscible if the reactive medium is basic [193], the combined reactor-decanter is a means to facilitate product separation and limit the formation of soap. Contactors combined with membranes, which retain unreacted glycerides, are also a recommended means for obtaining high purity products.

CONCLUSIONS

A review of the different process equipment that can be used to intensify FAME production has been presented. Continuous process technologies that intensify mixing and fluid contacted are recommended because the four reactions related to FAME production involved immiscible liquid–liquid reactants. The product separation step has been taken into account in the discussion and equipment that combine the reaction and separation steps are particularly recommended. The implementation of heterogeneous catalysis in these innovative process intensification technologies has also been discussed. Finally, specific process equipment has been recommended for the intensification of FAME based on different selection criteria.

It can be seen from this review that various types of innovative process equipment, such as cavitational reactors, oscillatory baffled reactors, microwave reactors, reactive distillation, static mixers and microstructured reactors enable significant mass transfer enhancement and improved performance of FAME production compared with conventional batch tank processes. Furthermore, the integration of continuous reaction and separation units appears to provide the best means for intensifying FAME production, leaving thus a vast number of configurations to be explored or invented. Amongst the different reactor types mentioned in this review, several are poorly understood in terms of physical phenomena and operating characteristics in liquid–liquid reaction applications. In particular, much fundamental knowledge on the operation of cavitational reactors, oscillatory baffled reactors and microwave reactors is still required. Further to this, the development of microwave reactors that are adapted to industrial FAME demands still requires significant work and considering this, continuous microwave reactors may be particularly interesting to explore. It should also be pointed out that heterogeneous catalysts have shown provide attractive results in terms of reaction performance in certain equipment (e.g. microwave reactors, reactive distillation) and therefore deserve to be explored further in other innovative process equipment, such as oscillatory baffled reactors and cavitational reactors.

Indeed, the choice of one process technology over another for FAME production is not that straightforward and it clearly depends on the global process objectives. Of course, some equipment (e.g. static mixers and microwave reactors) may be attractive due to their low energy consumption; others, such as reactive distillation, may be preferred due to effective product separation despite high energy requirements. However, although the number of studies on the feasibility of different process technologies for FAME production is significant, detailed economical and energy efficiency analyses of the various technologies and processes are still needed. These, in addition to studies on reaction performance, will of course be vital for the development of sustainable and green FAME production processes and therefore should be considered in the future studies.

ACKNOWLEDGMENTS

This study is part of the AGRIBTP Project on bioproducts for building and public work that is funded by the European Union, the French Government and the Région Midi-Pyrénées.

REFERENCES

1. J. Kansedo, K.T. Lee, S. Bhatia, Cerbera odollam (sea mango) oil as a promising non-edible feedstock for biodiesel production, Fuel 88 (2009) 1148–1150.
2. J.O. Metzger, Fats and oils as renewable feedstock for chemistry, European Journal of Lipid Science and Technology 111 (2009) 865–876.
3. M.M. Gui, K.T. Lee, S. Bhatia, Feasibility of edible oil vs. non-edible oil vs. waste edible oil as biodiesel feedstock, Energy 33 (2008) 1646–1653.

4.   H. Taher, S. Al-Zuhair, A.H. Al-Marzouqi, Y. Haik, M.M. Farid, A review of enzymatic transesterification of microalgal oil-based biodiesel using supercritical technology, Enzyme Research 2011 (2011) 1–25.

5.   Y. Zhang, M.A. Dube, D.D. McLean, M. Kates, Biodiesel production from waste cooking oil: 2. Economic assessment and sensitivity analysis, Bioresource Technology 90 (2003) 229–240.

6.   A. Banerjee, R. Chakraborty, Parametric sensitivity in transesterification of waste cooking oil for biodiesel production—a review, Resources, Conservation and Recycling 53 (2009) 490–497.

7.   M.G. Kulkarni, A.K. Dalai, Waste cooking oil an economical source for biodiesel: a review, Industrial & Engineering Chemistry Research 45 (2006) 2901–2913.

8.   Article R1331–2, n.d.

9.   Food Standards Agency website, (n.d.).

10.   Restaurant Technologies Inc. .

11.   Shakopee Mdewakanton Dakota Community. .

12.   D. Bégin, M. Gérin, I. de recherche en santé et en sécurité du travail du Québec, U. de M.D. de médecine du travail et d'hygiène du milieu, in: La substitution des solvants par la N-méthyl-2-pyrrolidone, Institut de recherche en santé et en sécurité du travail du Québec, 1999.

13.   Kooperationsstelle Hamburg, Layman Report – Reduction of VOC emissions by using fatty acid esters for metal cleaning processes, n.d.

14.   A.J. Hutchinson, V.G. Gomes, L.J. Hyde, Engineering an anti-graffiti system: a study in industrial product design, Chemical Engineering & Technology 27 (2004) 874–879.

15.   NDCEE, NDCEE Determines Lactate Esters Are Effective Nontoxic Cleaning Materials, 2003.

16.   R. von Wedel, Cytosol–cleaning oiled shorelines with a vegetable oil biosolvent, Spill Science & Technology Bulletin 6 (2000) 357–359.

17. P.A. Noirot, Green ink for all colors, Ink Maker 82 (2004) 29–31.

18. P. Van Broekhuizen, Technical and non-technical aspects in the substitution of mineral oil based products by vegetable alternatives, Agro Food Industry Hi-tech 12 (2001) 39–43.

19. R. Höfer, J. Bigorra, Green chemistry – a sustainable solution for industrial specialties applications, Green Chemistry 9 (2007) 203.

20. D. Charlemagne, The contribution of lipochemistry to the plant protection industry, OCL-Oléagineux, Corps Gras, Lipides 6 (1999) 401–404.

21. D.Y.C. Leung, X. Wu, M.K.H. Leung, A review on biodiesel production using catalyzed transesterification, Applied Energy 87 (2010) 1083–1095.

22. M.K. Lam, K.T. Lee, A.R. Mohamed, Homogeneous, heterogeneous and enzymatic catalysis for transesterification of high free fatty acid oil (waste cooking oil) to biodiesel: a review, Biotechnology Advances 28 (2010) 500– 518.

23. C.C. Enweremadu, M.M. Mbarawa, Technical aspects of production and analysis of biodiesel from used cooking oil—a review, Renewable and Sustainable Energy Reviews 13 (2009) 2205–2224.

24. A.P. Vyas, J.L. Verma, N. Subrahmanyam, A review on FAME production processes, Fuel 89 (2010) 1–9.

25. Z. Helwani, M.R. Othman, N. Aziz, W.J.N. Fernando, J. Kim, Technologies for production of biodiesel focusing on green catalytic techniques: a review, Fuel Processing Technology 90 (2009) 1502–1514.

26. E.M. Shahid, Y. Jamal, Production of biodiesel: a technical review, Renewable and Sustainable Energy Reviews (2011).

27. P.-L. Boey, G.P. Maniam, S.A. Hamid, Performance of calcium oxide as a heterogeneous catalyst in biodiesel production: a review, Chemical Engineering Journal 168 (2011) 15–22.

28. Y.C. Sharma, B. Singh, J. Korstad, Latest developments on application of heterogenous basic catalysts for an efficient

and eco friendly synthesis of biodiesel: a review, Fuel 90 (2011) 1309–1324.

29. A.P. Chouhan, A.K. Sarma, Modern heterogeneous catalysts for biodiesel production: a comprehensive review, Renewable and Sustainable Energy Reviews 15 (2011) 4378–4399.

30. S. Semwal, A.K. Arora, R.P. Badoni, D.K. Tuli, Biodiesel production using heterogeneous catalysts, Bioresource Technology 102 (2011) 2151–2161.

31. I.M. Atadashi, M.K. Aroua, A.R. Abdul Aziz, N.M.N. Sulaiman, The effects of catalysts in biodiesel production: a review, Journal of Industrial and Engineering Chemistry 19 (2013) 14–26.

32. Z. Qiu, L. Zhao, L. Weatherley, Process intensification technologies in continuous biodiesel production, Chemical Engineering and Processing: Process Intensification 49 (2010) 323–330.

33. A. Talebian-Kiakalaieh, N.A.S. Amin, H. Mazaheri, A review on novel processes of biodiesel production from waste cooking oil, Applied Energy 104 (2013) 683–710.

34. G.L. Maddikeri, A.B. Pandit, P.R. Gogate, Intensification approaches for biodiesel synthesis from waste cooking oil: a review, Industrial & Engineering Chemistry Research 51 (2012) 14610–14628.

35. P.P. Oh, H.L.N. Lau, J. Chen, M.F. Chong, Y.M. Choo, A review on conventional technologies and emerging process intensification (PI) methods for biodiesel production, Renewable and Sustainable Energy Reviews 16 (2012) 5131–5145.

36. A.A. Refaat, Different techniques for the production of biodiesel from waste vegetable oil, International Journal of Environmental Science and Technology 7 (2010) 183–213.

37. A.A. Kiss, C.S. Bildea, A review of biodiesel production by integrated reactive separation technologies, Journal of Chemical Technology & Biotechnology 87 (2012) 861–879.

38. I.M. Atadashi, M.K. Aroua, A.A. Aziz, Biodiesel separation and purification: a review, Renewable Energy 36 (2011) 437–443.

39. M. Canakci, J. Van Gerpen, Biodiesel production from oils and fats with high free fatty acids, Transactions-American Society of Agricultural Engineers 44 (2001) 1429–1436.

40. V. Makareviciene, E. Sendzikiene, P. Janulis, Solubility of multi-component biodiesel fuel systems, Bioresource Technology 96 (2005) 611–616.

41. J.M. Encinar, J.F. González, A. Rodríguez-Reinares, Biodiesel from used frying oil. Variables affecting the yields and characteristics of the biodiesel, Industrial & Engineering Chemistry Research 44 (2005) 5491–5499.

42. J. Cvengroš, Z. Cvengrošová, Used frying oils and fats and their utilization in the production of methyl esters of higher fatty acids, Biomass and Bioenergy 27 (2004) 173–181.

43. H. Fukuda, A. Kondo, H. Noda, Biodiesel fuel production by transesterification of oils, Journal of Bioscience and Bioengineering 92 (2001) 405–416.

44. A.A. Refaat, N.K. Attia, H.A. Sibak, S.T. El Sheltawy, G.I. El Diwani, et al., Production optimization and quality assessment of biodiesel from waste vegetable oil, International Journal of Environmental Science and Technology 5 (2008) 75–82.

45. B. Freedman, E.H. Pryde, T.L. Mounts, Variables affecting the yields of fatty esters from transesterified vegetable oils, Journal of the American Oil Chemists' Society 61 (1984) 1638–1643.

46. C.L. Chen, C.C. Huang, D.T. Tran, J.S. Chang, Biodiesel synthesis via heterogeneous catalysis using modified strontium oxides as the catalysts, Bioresource Technology 113 (2012) 8–13.

47. X. Liu, H. He, Y. Wang, S. Zhu, Transesterification of soybean oil to biodiesel using SrO as a solid base catalyst, Catalysis Communications 8 (2007) 1107– 1111.

48. E. Tur, B. Onal-Ulusoy, E. Akdogan, M. Mutlu, Surface modification of polyethersulfone membrane to improve its

hydrophobic characteristics for waste frying oil filtration: radio frequency plasma treatment, Journal of Applied Polymer Science 123 (2012) 3402–3411.

49. C.E.C. Rodrigues, C.B. Gonçalves, E. Batista, A.J.A. Meirelles, Deacidification of vegetable oils by solvent extraction, Recent Patents on Engineering 1 (2007) 95–102.

50. J.F. Gomes, D. Vergueiro, Study on the glycerolysis reaction of high free fatty acid oils for use as biodiesel feedstock, Fuel Processing Technology 92 (2011) 1225–1229.

51. D.A. Echeverri, F. Cardeño, L.A. Rios, Glycerolysis of soybean oil with crude glycerol containing residual alkaline catalysts from biodiesel production, Journal of the American Oil Chemists' Society 88 (2010) 551–557.

52. M. Kotwal, S.S. Deshpande, D. Srinivas, Esterification of fatty acids with glycerol over Fe–Zn double-metal cyanide catalyst, Catalysis Communications 12 (2011) 1302–1306.

53. Y. Wang, R.H. Natelson, L.F. Stikeleather, W.L. Roberts, Solid superacid catalyzed glycerol esterification of free fatty acids in waste cooking oil for biodiesel production, European Journal of Lipid Science and Technology (2012).

54. S.M. Kim, J.S. Rhee, Production of medium-chain glycerides by immobilized lipase in a solvent-free system, Journal of the American Oil Chemists' Society 68 (1991) 499–501.

55. A. Robles Medina, L. Esteban Cerdán, A. Giménez Giménez, B. Camacho Páez, M.J. Ibáñez González, E. Molina Grima, Lipase-catalyzed esterification of glycerol and polyunsaturated fatty acids from fish and microalgae oils, Progress in Industrial Microbiology 35 (1999) 379–391.

56. W.-C. Wang, R.H. Natelson, L.F. Stikeleather, W.L. Roberts, CFD simulation of transient stage of continuous countercurrent hydrolysis of canola oil, Computers & Chemical Engineering 43 (2012) 108–119.

57. H.L. Barnebey, A.C. Brown, Continuous fat splitting plants using the Colgate-Emery process, Journal of the American Oil Chemists' Society (1948) 95–99.

58. J.K. Satyarthi, D. Srinivas, P. Ratnasamy, Hydrolysis of vegetable oils and fats to fatty acids over solid acid catalysts, Applied Catalysis A: General 391 (2011) 427–435.

59. G.D. Machado, D.A.G. Aranda, M. Castier, V.F. Cabral, L. Cardozo-Filho, Computer simulations of fatty acid esterification in reactive distillation columns, Industrial & Engineering Chemistry Research 50 (2011) 10176– 10184.

60. H. Noureddini, D. Zhu, Kinetics of transesterification of soybean oil, Journal of the American Oil Chemists' Society 74 (1997) 1457–1463.

61. H.A. Farag, A. El-Maghraby, N.A. Taha, Optimization of factors affecting esterification of mixed oil with high percentage of free fatty acid, Fuel Processing Technology 92 (2011) 507– 510.

62. D. Ballerini, T. Chapus, G. Hillion, X. Montagne, Les esters d'huile végétale, in: Les Biocarburants, 2011: p. 129.

63. D. Frascari, M. Zuccaro, A. Paglianti, D. Pinelli, Optimization of mechanical agitation and evaluation of the mass-transfer resistance in the oil transesterification reaction for biodiesel production, Industrial & Engineering Chemistry Research 48 (2009) 7540–7549.

64. Y. Zhang, M.A. Dube, D.D. McLean, M. Kates, Biodiesel production from waste cooking oil: 1. Process design and technological assessment, Bioresource Technology 89 (2003) 1–16.

65. P. Cintas, S. Mantegna, E.C. Gaudino, G. Cravotto, A new pilot flow reactor for high-intensity ultrasound irradiation, Application to the Synthesis of Biodiesel, Ultrasonics Sonochemistry 17 (2010) 985–989.

66. M. Slinn, K. Kendall, Developing the reaction kinetics for a biodiesel reactor, Bioresource Technology 100 (2009) 2324– 2327.

67. G. Guan, M. Teshima, C. Sato, S. Mo Son, M. Faisal Irfan, K. Kusakabe, et al., Two-phase flow behavior in microtube

reactors during biodiesel production from waste cooking oil, AIChE Journal 56 (2009) 1383–1390.

68. O.S. Stamenkovic, Z.B. Todorovic, M.L. Lazic, V.B. Veljkovic, D.U. Skala, Kinetics of sunflower oil methanolysis at low temperatures, Bioresource Technology 99 (2008) 1131–1140.

69. R. Heryanto, M. Hasan, E.C. Abdullaha, A.C. Kumoro, Solubility of stearic acid in various organic solvents and its prediction using non-ideal solution models 33 (2007) 469-472.

70. L. Marrone, L. Pasco, D. Moscatelli, S. Gelosa, Liquid-liquid phase equilibrium in glycerol-methanol-fatty acids systems, Chemical Engineering 11 (2007).

71. D.A. Aranda, C. da Silva, C. Detoni, Current processes in brazilian biodiesel production, International Review of Chemical Engineering 1 (2009) 603–608.

72. D. Reay, C. Ramshaw, A. Harvey, Process Intensification: Engineering for Efficiency, Sustainability and Flexibility, first ed., Butterworth-Heineman, 2008.

73. M. Poux, P. Cognet, C. Gourdon, Génie des procédés durables: Du concept à la concrétisation industrielle, Paris, 2010.

74. W. Ehrfeld, V. Hessel, H. Löwe, Microreactors: New Technology for Modern Chemistry, Wiley-VCH, Weinheim, 2000.

75. N. Kockmann, Micro Process Engineering: Fundamentals, Devices, Fabrication and Applications, Wiley-VCH, Weinheim, 2006.

76. M.N. Kashid, L. Kiwi-Minsker, Microstructured reactors for multiphase reactions: state of the art, Industrial & Engineering Chemistry Research 48 (2009) 6465–6485.

77. A.L. Dessimoz, L. Cavin, A. Renken, L. Kiwi-Minsker, Liquid–liquid two-phase flow patterns and mass transfer characteristics in rectangular glass microreactors, Chemical Engineering Science 63 (2008) 4035–4044.

78. J. Jovanovic, E. Rebrov, T.A. Nijhuis, M.T. Kreutzer, V. Hessel, J.C. Schouten, Liquid-liquid flow in long capillaries: hydrodynamic flow patterns and extraction performance, Industrial & Engineering Chemistry Research 51 (2011) 1015–1026.

79. A. Pohar, M. Lakner, I. Plazl, Parallel flow of immiscible liquids in a microreactor: modeling and experimental study, Microfluidics and Nanofluidics 12 (2011) 307–316.

80. Z. Wen, X. Yu, S.-T. Tu, J. Yan, E. Dahlquist, Intensification of biodiesel synthesis using zigzag micro-channel reactors, Bioresource Technology 100 (2009) 3054–3060.

81. G. Guan, K. Kusakabe, K. Moriyama, N. Sakurai, Transesterification of sunflower oil with methanol in a microtube reactor, Industrial & Engineering Chemistry Research 48 (2009) 1357–1363.

82. J. Sun, J. Ju, L. Ji, L. Zhang, N. Xu, Synthesis of biodiesel in capillary microreactors, Industrial & Engineering Chemistry Research 47 (2008) 1398–1403.

83. R. Jachuck, G. Pherwani, S.M. Gorton, Green engineering: continuous production of biodiesel using an alkaline catalyst in an intensified narrow channel reactor, Journal of Environmental Monitoring 11 (2009) 642.

84. E.E. Kalu, K.S. Chen, T. Gedris, Continuous-flow biodiesel production using slit-channel reactors, Bioresource Technology 102 (2011) 4456–4461.

85. D.E. Leng, R.V. Calabrese, E.L. Paul, V.A. Atiemo-Obeng, S.M. Kresta, Immiscible Liquid-Liquid Systems in Handbook of Industrial Mixing: Science & Practice, John Wiley & Sons Inc., Hoboken NJ, 2004.

86. P. Sun, J. Sun, J. Yao, L. Zhang, N. Xu, Continuous production of biodiesel from high acid value oils in microstructured reactor by acid-catalyzed reactions, Chemical Engineering Journal 162 (2010) 364–370.

87. S. Matsuura, R. Ishii, T. Itoh, S. Hamakawa, T. Tsunoda, T. Hanaoka, et al., Immobilization of enzyme-encapsulated nanoporous material in a microreactor and reaction analysis, Chemical Engineering Journal 167 (2011) 744–749.

88. Corning Microreactor technology brings new market to Corning, (n.d.).

89. Chart: Compact Heat Exchange Reactors, (n.d.).

90. Ehrfeld website, (n.d.).

91. IMM Website: Liquid/Liquid Microreactor, (n.d.).

92. P.R. Gogate, Cavitational reactors for process intensification of chemical processing applications: a critical review, Chemical Engineering and Processing: Process Intensification 47 (2008) 515–527.

93. V.L. Gole, P.R. Gogate, A review on intensification of synthesis of biodiesel from sustainable feed stock using sonochemical reactors, Chemical Engineering and Processing: Process Intensification 53 (2012) 1–9.

94. P.R. Gogate, R.K. Tayal, A.B. Pandit, Cavitation: a technology on the horizon, Current Science 91 (2006) 35–46.

95. V.B. Veljkovic, J.M. Avramovic, O.S. Stamenkovic, Biodiesel production by ultrasound-assisted transesterification: state of the art and the perspectives, Renewable and Sustainable Energy Reviews 16 (2012) 1193–1209.

96. A.S. Badday, A.Z. Abdullah, K.T. Lee, M.S. Khayoon, Intensification of biodiesel production via ultrasonic-assisted process: a critical review on fundamentals and recent development, Renewable and Sustainable Energy Reviews 16 (2012) 4574–4587.

97. K. Ramachandran, T. Suganya, N. Nagendra Gandhi, S. Renganathan, Recent developments for biodiesel production by ultrasonic assist transesterification using different heterogeneous catalyst: a review, Renewable and Sustainable Energy Reviews 22 (2013) 410–418.

98. C. Stavarache, M. Vinatoru, R. Nishimura, Y. Maeda, Conversion of vegetable oil to biodiesel using ultrasonic irradiation, Cheminfor 34 (2003).

99. C. Stavarache, M. Vinatoru, R. Nishimura, Y. Maeda, Fatty acids methyl esters from vegetable oil by means of ultrasonic energy, Ultrasonics Sonochemistry 12 (2005) 367–372.

100. X. Fan, F. Chen, X. Wang, Ultrasound-assisted synthesis of biodiesel from crude cottonseed oil using response surface methodology, Journal of Oleo Science 59 (2010) 235–241.

101. J.A. Colucci, E.E. Borrero, F. Alape, Biodiesel from an alkaline transesterification reaction of soybean oil using ultrasonic mixing, Journal of the American Oil Chemists' Society 82 (2005) 525–530.

102. A.K. Singh, S.D. Fernando, R. Hernandez, Base-catalyzed fast transesterification of soybean oil using ultrasonication, Energy & Fuels 21 (2007) 1161–1164.

103. P. Chand, V.R. Chintareddy, J.G. Verkade, D. Grewell, Enhancing biodiesel production from soybean oil using ultrasonics, Energy & Fuels 24 (2010) 2010–2015.

104. H.D. Hanh, N. The Dong, C. Stavarache, K. Okitsu, Y. Maeda, R. Nishimura, Methanolysis of triolein by low frequency ultrasonic irradiation, Energy Conversion and Management 49 (2008) 276–280.

105. L.S. Teixeira, J.C. Assis, D.R. Mendonça, I.T. Santos, P.R. Guimarães, L.A. Pontes, et al., Comparison between conventional and ultrasonic preparation of beef tallow biodiesel, Fuel Processing Technology 90 (2009) 1164–1166.

106. L.T. Thanh, K. Okitsu, Y. Sadanaga, N. Takenaka, Y. Maeda, H. Bandow, Ultrasound-assisted production of biodiesel fuel from vegetable oils in a small scale circulation process, Bioresource Technology 101 (2010) 639–645.

107. L.T. Thanh, K. Okitsu, Y. Sadanaga, N. Takenaka, Y. Maeda, H. Bandow, A twostep continuous ultrasound assisted production of biodiesel fuel from waste cooking oils: a practical and

economical approach to produce high quality biodiesel fuel, Bioresource Technology 101 (2010) 5394–5401.

108. C. Stavarache, M. Vinatoru, Y. Maeda, H. Bandow, Ultrasonically driven continuous process for vegetable oil transesterification, Ultrasonics Sonochemistry 14 (2007) 413–417.

109. M.A. Kelkar, P.R. Gogate, A.B. Pandit, Intensification of esterification of acids for synthesis of biodiesel using acoustic and hydrodynamic cavitation, Ultrasonics Sonochemistry 15 (2008) 188–194.

110. V.G. Deshmane, P.R. Gogate, A.B. Pandit, Process intensification of synthesis process for medium chain glycerides using cavitation, Chemical Engineering Journal 145 (2008) 351–354.

111. A.B. Pandit, J.B. Joshi, Hydrolysis of fatty oils: effect of cavitation, Chemical Engineering Science 48 (1993) 3440–3442.

112. O.V. Kozyuk, Apparatus and Method for Producing Biodiesel from Fatty Acid Feedstock, US Patent 7754905, 2010.

113. D. Ghayal, A.B. Pandit, V.K. Rathod, Optimization of biodiesel production in a hydrodynamic cavitation reactor using used frying oil, Ultrasonics Sonochemistry 20 (2013) 322–328.

114. P.R. Muniyappa, S.C. Brammer, H. Noureddini, Improved conversion of plant oils and animal fats into biodiesel and co-product, Chemical and Biomolecular Engineering Research and Publications (1996).

115. Arisdyne website, (n.d.).

116. Hydro Dynamics, Inc. website, (n.d.).

117. D.G. Mancosky, D.A. Armstead, T. McGurk, G. Hopkins, K. Hudson, The Use of a Controlled Cavitation Reactor for Bio-Diesel Production, 2007.

118. D.M.P. Mingos, D.R. Baghurst, Tilden lecture. Applications of microwave dielectric heating effects to synthetic problems in chemistry, Chemical Society Reviews 20 (1991) 1.

119. P. Lidström, J. Tierney, B. Wathey, J. Westman, Microwave assisted organic synthesis–a review, Tetrahedron 57 (2001) 9225–9283.

120. A. Loupy, Microwave en Organic Synthesis, Wiley-VCH, Weinheim, 2006.

121. C.O. Kappe, D. Dallinger, S.S. Murphree, Practical Microwave Synthesis for Organic Chemists: Strategies, Instruments and Protocols, Wiley-VCH, Weinheim, 2009.

122. F. Motasemi, F.N. Ani, A review on microwave-assisted production of biodiesel, Renewable and Sustainable Energy Reviews 16 (2012) 4719–4733.

123. V.G. Gude, P. Patil, E. Martinez-Guerra, S. Deng, N. Nirmalakhandan, Microwave energy potential for biodiesel production, Microwave Energy Potential for Biodiesel Production 1 (2013) 1–31.

124. B.G. Terigar, S. Balasubramanian, M. Lima, D. Boldor, Transesterification of soybean and rice bran oil with ethanol in a continuous-flow microwaveassisted system: yields, Quality, and Reaction Kinetics, Energy & Fuels 24 (2010) 6609–6615.

125. Y. Asamuka, Y. Ogawa, K. Maeda, K. Fukui, H. Kuramochi, Effects of microwave irradiation on triglyceride transesterification: experimental and theoretical studies, Biochemical Engineering Journal 58–59 (2011) 20–24.

126. A. Breccia, B. Esposito, G. Breccia Fratadocchi, A. Fini, Reaction between methanol and commercial seed oils under microwave irradiation, International Microwave Power Institute 34 (1999) 1–8.

127. N.E. Leadbeater, L.M. Stencel, Fast, easy preparation of biodiesel using microwave heating, Energy & Fuels 20 (2006) 2281–2283.

128. A.A. Refaat, S.T. El Sheltawy, K.U. Sadek, et al., Optimum reaction time, performance and exhaust emissions of biodiesel produced by microwave irradiation, International Journal of Environmental Science and Technology 5 (2008) 315–322.

129. N. Azcan, A. Danisman, Microwave assisted transesterification of rapeseed oil, Fuel 87 (2008) 1781–1788.

130. Y. Zu, S. Zhang, Y. Fu, W. Liu, Z. Liu, M. Luo, et al., Rapid microwave-assisted transesterification for the preparation of fatty acid methyl esters from the oil of yellow horn (Xanthoceras sorbifolia Bunge.), European Food Research and, Technology 229 (2009) 43–49.

131. A. Kanitkar, S. Balasubramanian, M. Lima, D. Boldor, A critical comparison of methyl and ethyl esters production from soybean and rice bran oil in the presence of microwaves, Bioresource Technology 102 (2011) 7896–7902.

132. M.Z. Duz, A. Saydut, G. Ozturk, Alkali catalyzed transesterification of safflower seed oil assisted by microwave irradiation, Fuel Processing Technology 92 (2011) 308–313.

133. V. Lertsathapornsuk, R. Pairintra, K. Aryusuk, K. Krisnangkura, Microwave assisted in continuous biodiesel production from waste frying palm oil and its performance in a 100 kW diesel generator, Fuel Processing Technology 89 (2008) 1330–1336.

134. T.M. Barnard, N.E. Leadbeater, M.B. Boucher, L.M. Stencel, B.A. Wilhite, Continuous-flow preparation of biodiesel using microwave heating, Energy & Fuels 21 (2007) 1777–1781.

135. A.M. Socha, J.K. Sello, Efficient conversion of triacylglycerols and fatty acids to biodiesel in a microwave reactor using metal triflate catalysts, Organic & Biomolecular Chemistry 8 (2010) 4753.

136. S. Zhang, Y.G. Zu, Y.J. Fu, M. Luo, D.Y. Zhang, T. Efferth, Rapid microwaveassisted transesterification of yellow horn oil to biodiesel using a heteropolyacid solid catalyst, Bioresource Technology 101 (2010) 931–936.

137. M. Koberg, R. Abu-Much, A. Gedanken, Optimization of bio-diesel production from soybean and wastes of cooked oil: combining dielectric microwave irradiation and a SrO catalyst, Bioresource Technology 102 (2011) 1073– 1078.

138. P. Khemthong, C. Luadthong, W. Nualpaeng, P. Changsuwan, P. Tongprem, N. Viriya-empikul, et al., Industrial eggshell

wastes as the heterogeneous catalysts for microwave-assisted biodiesel production, Catalysis Today 190 (2012) 112–116.

139. C.A.R. Melo Ju´ nior, C.E.R. Albuquerque, J.S.A. Carneiro, C. Dariva, M. Fortuny, A.F. Santos, et al., Solid-acid-catalyzed esterification of oleic acid assisted by microwave heating, Industrial & Engineering Chemistry Research 49 (2010) 12135–12139.

140. C.A.R. Melo-Ju´ nior, C.E.R. Albuquerque, M. Fortuny, C. Dariva, S. Egues, A.F. Santos, et al., Use of microwave irradiation in the noncatalytic esterification of C18 fatty acids, Energy & Fuels 23 (2009) 580–585.

141. D. Kim, J. Choi, G.J. Kim, S.K. Seol, S. Jung, Accelerated esterification of free fatty acid using pulsed microwaves, Bioresource Technology 102 (2011) 7229–7231.

142. D. Kim, J. Choi, G.J. Kim, S.K. Seol, Y.C. Ha, M. Vijayan, et al., Microwaveaccelerated energy-efficient esterification of free fatty acid with a heterogeneous catalyst, Bioresource Technology 102 (2011) 3639–3641.

143. R. Luque, J. Budarin, J.H. Clark, D.J. Macquarrie, Glycerol transformations on polysaccharide derived mesoporous materials, Applied Catalysis B: Environmental 82 (2008) 157–162.

144. S.F. Marcel, K.J. Lie, Y.-K. Cheung, The use of a microwave oven in the chemical transformation of long chain fatty acid esters, Lipids (1988) 367– 369.

145. R.K. Saxena, J. Isar, S. Saran, R. Kaushik, W.S. Davidson, Efficient microwaveassisted hydrolysis of triolein and synthesis of bioester, bio-surfactant and triglycerides using Aspergillus carneus lipase, Current Science 89 (2005) 1000–1003.

146. C.S. Fang, P.M.C. Lai, Microwave heating and separation of water-in-oil emulsions, Journal of Microwave Power and Electromagnetic Energy 30 (1995) 46–57.

147. A.P. Harvey, M.R. Mackley, T. Seliger, Process intensification of biodiesel production using a continuous oscillatory flow

reactor, Journal of Chemical Technology & Biotechnology 78 (2003) 338–341.

148. A.N. Phan, A.P. Harvey, M. Rawcliffe, Continuous screening of base-catalysed biodiesel production using new designs of mesoscale oscillatory baffled reactors, Fuel Processing Technology 92 (2011) 1560–1567.

149. M. Zheng, R.L. Skelton, M.R. Mackley, Biodiesel reaction screening using oscillatory flow meso reactors, Process Safety and Environmental Protection 85 (2007) 365–371.

150. http://www.nitechsolutions.co.uk/market-sectors/clean-technology/, (n.d.).

151. A.W. Etchells III, C.F. Meyer, Chapter 7: mixing in pipelines, in: Handbook of Industrial Mixing: Science and Practice, John Wiley & Sons, 2004.

152. C. Xuereb, M. Poux, J. Bertrand, Mélangeurs statiques, in: Agitation et Mélange, Dunod, 2006.

153. A. Baumann, S.A.K. Jeelani, B. Holenstein, P. Stössel, E.J. Windhab, Flow regimes and drop break-up in SMX and packed bed static mixers, Chemical Engineering Science 73 (2012) 354–365.

154. X. Yu, Z. Wen, Y. Lin, S.T. Tu, Z. Wang, J. Yan, Intensification of biodiesel synthesis using metal foam reactors, Fuel 89 (2010) 3450–3456.

155. M. Al-Atabi, Design and assessment of a novel static mixer, The Canadian Journal of Chemical Engineering 89 (2011) 550–554.

156. D. Frascari, M. Zuccaro, D. Pinelli, A. Paglianti, A pilot-scale study of alkalicatalyzed sunflower oil transesterification with static mixing and with mechanical agitation, Energy & Fuels 22 (2008) 1493–1501.

157. J.C. Thompson, B.B. He, Biodiesel production using static mixers, Transactions of the ASABE 50 (2007) 161–165.

158. M.B. Boucher, C. Weed, N.E. Leadbeater, B.A. Wilhite, J.D. Stuart, R.S. Parnas, Pilot scale two-phase continuous flow

biodiesel production via novel laminar flow reactorseparator, Energy & Fuels 23 (2009) 2750–2756.

159. S.A. Unker, M.B. Boucher, K.R. Hawley, A.A. Midgette, J.D. Stuart, R.S. Parnas, Investigation into the relationship between the gravity vector and the flow vector to improve performance in two-phase continuous flow biodiesel reactor, Bioresource Technology 101 (2010) 7389–7396.

160. E. Santacesaria, M. Di Serio, R. Tesser, M. Tortorelli, R. Turco, V. Russo, A simple device to test biodiesel process intensification, Chemical Engineering and Processing: Process Intensification 50 (2011) 1085–1094.

161. E. Santacesaria, M. Di Serio, R. Tesser, R. Turco, M. Tortorelli, V. Russo, Biodiesel process intensification in a very simple microchannel device, Chemical Engineering and Processing: Process Intensification 52 (2012) 47– 54.

162. J.L. Massingill, P.N. Patel, M. Guntupalli, C. Garret, C. Ji, High efficiency nondispersive reactor for two-phase reactions, Organic Process Research & Development 12 (2008) 771– 777.

163. M. Yoshimune, K. Haraya, Microporous carbon membranes, in: Membranes for Membrane Reactors: Preparation, Optimization and Selection, Wiley, 2011.

164. X. Feng, R.Y.M. Huang, Liquid separation by membrane pervaporation: a review, Industrial & Engineering Chemistry Research 36 (1997) 1048–1066.

165. S.H. Shuit, Y.T. Ong, K.T. Lee, B. Subhash, S.H. Tan, Membrane technology as a promising alternative in biodiesel production: a review, Biotechnology Advances 30 (2012) 1364–1380.

166. M.A. Dube, A.Y. Tremblay, J. Liu, Biodiesel production using a membrane reactor, Bioresource Technology 98 (2007) 639– 647.

167. P. Cao, A.Y. Tremblay, M.A. Dubé, K. Morse, Effect of membrane pore size on the performance of a membrane reactor for biodiesel production, Industrial & Engineering Chemistry Research 46 (2007) 52–58.

168. P. Cao, A.Y. Tremblay, M.A. Dube´, Kinetics of canola oil transesterification in a membrane reactor, Industrial & Engineering Chemistry Research 48 (2009) 2533–2541.

169. N. Sdrula, A study using classical or membrane separation in the biodiesel process, Desalination 250 (2010) 1070–1072.

170. J. Saleh, M.A. Dubé, A.Y. Tremblay, Separation of glycerol from FAME using ceramic membranes, Fuel Processing Technology 92 (2011) 1305–1310.

171. M.C.S. Gomes, P.A. Arroyo, N.C. Pereira, Biodiesel production from degummed soybean oil and glycerol removal using ceramic membrane, Journal of Membrane Science 378 (2011) 453–461.

172. H. Falahati, A.Y. Tremblay, The effect of flux and residence time in the production of biodiesel from various feedstocks using a membrane reactor, Fuel 91 (2012) 126–133.

173. B. Sarkar, S. Sridhar, K. Saravanan, V. Kale, Preparation of fatty acid methyl ester through temperature gradient driven pervaporation process, Chemical Engineering Journal 162 (2010) 609–615.

174. W. Shi, B. He, J. Ding, J. Li, F. Yan, X. Liang, Preparation and characterization of the organic–inorganic hybrid membrane for biodiesel production, Bioresource Technology 101 (2010) 1501–1505.

175. M. Zhu, B. He, W. Shi, Y. Feng, J. Ding, J. Li, et al., Preparation and characterization of PSSA/PVA catalytic membrane for biodiesel production, Fuel 89 (2010) 2299–2304.

176. L. Guerreiro, J.E. Castanheiro, I.M. Fonseca, R.M. Martin-Aranda, A.M. Ramos, J. Vital, Transesterification of soybean oil over sulfonic acid functionalised polymeric membranes, Catalysis Today 118 (2006) 166–171.

177. V.R. Murty, J. Bhat, P.K.A. Muniswaran, Hydrolysis of oils by using immobilized lipase enzyme: a review, Biotechnology and Bioprocess Engineering 7 (2002) 57–66.

178. G.J. Harmsen, Reactive distillation: the front-runner of industrial process intensification: a full review of commercial

applications, research, scale-up, design and operation, Chemical Engineering and Processing 46 (2007) 774– 780.

179. Ö. Yildirim, A.A. Kiss, E.Y. Kenig, Dividing wall columns in chemical process industry: a review on current activities, Separation and Purification Technology 80 (2011) 403–417.

180. B.B. He, A.P. Singh, J.C. Thompson, A novel continuous-flow reactor using reactive distillation for biodiesel production, Transactions of the ASAE 49 (2006) 107–112.

181. S. Steinigeweg, J. Gmehling, Esterification of a fatty acid by reactive distillation, Industrial & Engineering Chemistry Research 42 (2003) 3612– 3619.

182. A.A. Kiss, J.G. Segovia-Hernández, C.S. Bildea, E.Y. Miranda-Galindo, S. Hernández, Reactive DWC leading the way to FAME and fortune, Fuel (2012) 352–359.

183. A.P. Singh, J.C. Thompson, B.B. He, A continuous-flow reactive distillation reactor for biodiesel preparation from seed oils, in: Canada, 2004.

184. A.A. Kiss, F. Omota, A.C. Dimian, G. Rothenberg, The heterogeneous advantage: biodiesel by catalytic reactive distillation, Topics in Catalysis 40 (2006) 141–150.

185. A.C. Dimian, C.S. Bildea, F. Omota, A.A. Kiss, Innovative process for fatty acid esters by dual reactive distillation, Computers & Chemical Engineering 33 (2009) 743–750.

186. A.A. Kiss, C.S. Bildea, Integrated reactive absorption process for synthesis of fatty esters, Bioresource Technology 102 (2011) 490–498.

187. N. Asprion, G. Kaibel, Dividing wall columns: fundamentals and recent advances, Chemical Engineering and Processing: Process Intensification 49 (2010) 139–146.

188. S. Phuenduang, P. Siricharnsakunchai, L. Simasatitkul, W. Paengjuntuek, A. Arpornwichanop, in:Optimization of Biodiesel Production from Jatropha Oil Using Reactive Distillation, 2011.

189. N.G. Patil, A.I.G. Hermans, F. Benaskar, J. Meuldijk, L.A. Hulshof, V. Hessel, et al., Energy efficient and controlled flow

processing under microwave heating by using a millireactor-heat exchanger, AIChE Journal (2011).

190. M.E. Fabiyi, R.L. Skelton, Photocatalytic mineralisation of methylene blue using buoyant TiO2-coated polystyrene beads, Journal of Photochemistry and Photobiology A: Chemistry 132 (2000) 121–128.

191. J.G. Khinast, A. Bauer, D. Bolz, A. Panarello, Mass-transfer enhancement by static mixers in a wall-coated catalytic reactor, Chemical Engineering Science 58 (2003) 1063–1070.

192. R.B.R. Choudhury, The preparation and purification of monoglycerides. II. Direct esterification of fatty acids with glycerol, Journal of the American Oil Chemists' Society 39 (1962) 345–347.

Citations

CHAPTER 1

Denise Przybylski, Thore Rohwerder, Hauke Harms, and Roland H Mueller, Third-generation Feed Stocks for the Clean and Sustainable Biotechnological Production of Bulk Chemicals: Synthesis of 2-hydroxyisobutyric Acid, doi: 10.1186/2192-0567-2-11.

CHAPTER 2

Onwughara Innocent Nkwachukwu, Chukwu Henry Chima, Alaekwe Obiora Ikenna, and Lackson Albert, Focus on Potential Environmental Issues on Plastic World towards a Sustainable Plastic Recycling in Developing Countries, doi: 10.1186/2228-5547-4-34.

CHAPTER 3

Stijn W.H. Van Hulle, Helge J.P. Vandeweyer, Boudewijn D. Meesschaert, Peter A. Vanrolleghem, Pascal Dejans, Ann Dumoulin, Engineering aspects and practical application of autotrophic nitrogen removal from nitrogen rich streams, Chemical Engineering Journal, Volume 162, Issue 1, 1 August 2010, Pages 1-20, ISSN 1385-8947, http://dx.doi.org/10.1016/j.cej.2010.05.037.

CHAPTER 4

Santanu Bandyopadhyay, Targeting minimum waste treatment flow rate, Chemical Engineering Journal, Volume 152, Issues 2–3, 15 October 2009, Pages 367-375, ISSN 1385-8947, http://dx.doi.org/10.1016/j.cej.2009.04.060.

CHAPTER 5

Alex Mazubert, Martine Poux, Joëlle Aubin, Intensified processes for FAME production from waste cooking oil: A technological review, Chemical Engineering Journal, Volume 233, November 2013, Pages 201-223, ISSN 1385-8947, http://dx.doi.org/10.1016/j.cej.2013.07.063.

Index